MEMBRANES, IONS, AND IMPULSES

FASEB MONOGRAPHS

General Editor: KARL F. HEUMANN

MEMBRANES, IONS, AND IMPULSES

Edited by
John W. Moore
Duke University Medical Center

FASEB, Bethesda
PLENUM PRESS, New York and London

Library of Congress Cataloging in Publication Data

Main entry under title:

Membranes, ions, and impulses.

(FASEB monographs; v. 5)
"Colloquium sponsored by the American Physiological Society at the
58th annual meeting of the Federation of American Societies for Experi-
mental Biology, Atlantic City, N. J., April 9-11, 1974."
"Originally appeared in Federation proceedings, vol. 34, No. 5, 1975."
Includes index.
    1. Neural transmission—Congresses. 2. Cell membranes—Congresses. I.
Moore, John Wilson, 1920-   II. American Physiological Society
(Founded 1887) III. Federation of American Societies for Experimental
Biology. Federation proceedings. IV. Series: Federation of American
Societies for Experimental Biology. FASEB monographs; v. 5 [DNLM:
1. Cell membrane—Congresses. 2. Electric conductivity—Congresses. 3.
Membrane potentials—Congresses. W1 F202 v. 5/QT34 C714m 1974]
QP363.M52 1976   591.1'88   76-13841

ISBN 978-1-4684-2639-7   ISBN 978-1-4684-2637-3 (eBook)
DOI 10.1007/978-1-4684-2637-3

The material in this book originally appeared in *Federation
Proceedings* Vol. 34, No. 5, April 1975. First published in the
present form by Plenum Publishing Corporation in 1976.

Copyright ©1975, by the Federation of American Societies
Softcover reprint of the hardcover 1st edition 1975

for Experimental Biology

Plenum Press, New York is a Division of Plenum Publishing Corporation
227 West 17th Street, New York, N.Y. 10011

Contents

MEMBRANES AND GEOMETRICAL FACTORS—MUSCLE

MEMBRANES AND GEOMETRICAL FACTORS—NEURONS

MEMBRANES, IONS, AND IMPULSES

Kenneth S. Cole

Membranes, ions, and impulses

Dedication to Kacy[1] Cole

JOHN W. MOORE

Department of Physiology and Pharmacology
Duke University Medical Center, Durham, North Carolina 27710

The title of the Colloquium on *Membranes, Ions, and Impulses* was chosen from Cole's book, and it is a pleasure to dedicate this collection of papers given at the Colloquium to Kacy. It is a privilege to be in the unique position of having been associated with both Kacy Cole and Dan Tosteson, our American Physiological Society President for 1974.

I am grateful to Dan Tosteson for the opportunity to be Chairman and the American Physiological Society representative for the committee to plan this intersociety colloquium. I want to express appreciation for the enthusiastic work of the other members of the Planning Committee: Clay Armstrong—representing the Biophysical Society, Richard Podolsky—representing the Society of General Physiology, Wilfred Rall—representing the Neuroscience Society, and Arnold Schwartz—representing the American Society of Pharmacology and Experimental Therapeutics.

Kacy earned his Bachelor's degree from Oberlin College and his Ph.D. in Physics under Richtmeyer at Cornell University. He spent 2 years at Harvard as a National Research Council Fellow and 1 year at Leipzig with Peter Debye. He then took an appointment in the Department of Physiology at Columbia's College of Physicians and Surgeons— probably the first physicist ever appointed to such a position. In 1942 he began war related research at the University of Chicago's Metallurgical Lab, and in 1946 transferred to the Biophysics Department. In 1949 he became Scientific Director of the Naval Medical Research Institute. I joined him there and the brief excursions that we made to the Marine Biological Laboratory (MBL) in Woods Hole, Massachusetts, for summer research whetted his appetite to return to the laboratory bench. He looked across Bethesda's Wisconsin Avenue and found a position at the National Institutes of Health as Chief of the Laboratory of Biophysics (NINDB) with much less administrative responsibility. This allowed Kacy time to work on

---

[1] Widely used affectionate nickname for Kenneth S. Cole.

1

science again and to revive his previous custom of spending summers at Woods Hole working on squid axons.

That Kacy was truly at home at the MBL, was loudly attested to by the return of his constant and very deep humming. No one, to my knowledge, ever recognized a tune but heard a spontaneous and continuing development of what we must call Cole's Opus No. 1 *Themes and Variations for Human Vocal Cords.* There were memorable evenings spent at the Cap't. Kidd drinking beer and spinning yarns about his years as an able seaman aboard a Great Lakes freighter or about sailboat racing at Woods Hole.

Kacy's scientific career may be characterized by "3 C's": *Concepts*—as exemplified by the elegance and power of the concept of the voltage clamp; *Care and Concern* for execution of experiments—for example, establishing that the axon membrane could have a uniform potential difference where it surrounded a carefully treated axial wire; and *Correction* of data for errors—as exemplified by corrections for nonideal electrode geometry.

Kacy has exerted an enormous influence on membrane research in a variety of ways such as: *1*) writing many articles as well as his book *Membranes, Ions and Impulses* from which the title of this Colloquium was taken; *2*) lectures both here and abroad; *3*) personal guidance of many within and without his laboratory; and *4*) through the MBL Training Program on Excitable Membranes.

The extent of Kacy's influence is indicated by the breadth of the subject matter covered in this Colloquium, and it is a distinct pleasure to participate in its planning and dedicate it to such a very unique and special person.

An essential ionized acid group in sodium channels[1]

BERTIL HILLE

Department of Physiology and Biophysics, University of Washington School of Medicine, Seattle, Washington 98195

ABSTRACT

Several recent experiments demonstrate the presence of an essential negatively charged acid group within sodium channels. Sodium permeability titrates away at low pH as if controlled by an acid with a voltage-dependent apparent pK_a in the range between 5 and 6. The alkali ion permeability sequence of the channel is best explained by interactions between the cations and a strong negative charge in the channel. Block of sodium currents by a variety of metal and organic cations again points to a cation-coordinating site in the channel. The "blocking cations" and protons also oppose the binding of tetrodotoxin and saxitoxin. The negative charge in the channel seems to be essential in selecting appropriate cations and in lowering their activation energy for permeation. The same charge seems to form part of the toxin receptor. At present this charged group is the only chemical group known to be associated with sodium channels.—HILLE, B. An essential ionized acid group in sodium channels. *Federation Proc.* 34: 1318–1321, 1975.

O ver the last 6 years several types of experiments have led to the conclusion that there is an essential negatively charged group within the Na channel of excitable membranes. This charged group seems to be important in lowering the free energy required to remove water molecules from ions as they pass through the narrow selectivity filter of the channel and seems also to be part of the receptor for several toxins at the channel. Most of the evidence comes from work with single myelinated nerve fibers, but some work with unmye-linated nerves shows that the charged group is found more generally.

PROTONS BLOCK Na CHANNELS

In 1967 Drs. Camougis, Takman, and Tasse told me of their then unpublished observation that the specific Na channel poison tetrodotoxin (TTX) blocks nerves less well at high pH than at neutral pH, presumably meaning that the zwitterionic form

[1] Supported by grant NS 08174 from the National Institutes of Health.

3

of TTX is less active than the cation (2). Using the voltage clamp technique on single myelinated nerve fibers, I confirmed this result with TTX and with saxitoxin (STX) as well (15). As a control to these experiments, it was also necessary to investigate the effects of external pH changes without toxin (16). The seminal observation came from these control experiments. Alkaline solutions did little to the voltage clamp properties of nodes of Ranvier, but acidic solutions reduced Na currents strongly. Indeed the currents were changed at low pH in two ways. First the peak Na permeability-voltage curve (P_{Na}-E curve) was shifted to more positive voltages as if the calcium concentration had been raised (9). Second P_{Na} was reduced at all voltages as if some Na channels were reversibly blocked by protons. The reduction of P_{Na} was fitted by supposing that a single negative charge with an apparent pK_a of 5.2 controls the permeability of each Na channel as in Fig. 1. The group must be ionized for the channel to function. It was suggested that the negative

charge might also be part of the receptor for TTX, STX, and local anesthetic molecules acting in the cationic form. The experiments were confirmed in myelinated nerve by Drouin and The (6), but there are no detailed studies on the effects of low pH on Na channels of other tissues.

In 1969 Ann Woodhull and I began to look for interactions between voltage shifts produced by changes in external calcium concentration and pH (31). She recognized that the description of low pH effects as a voltage shift of the peak P_{Na}-E curve and a superimposed block of a given percentage of channels was inadequate. As can be seen in Fig. 2, the peak P_{Na}-E curve at pH 5 did not flatten out at high depolarizations as at pH 7 but rather continued to rise, implying that the block of channels by protons depends on membrane potential. Fewer channels are blocked for large depolarizations than for small depolarizations as if depolarization drives protons out of their binding site. This voltage dependence was quantitatively accounted for by assuming that the protonatable site is

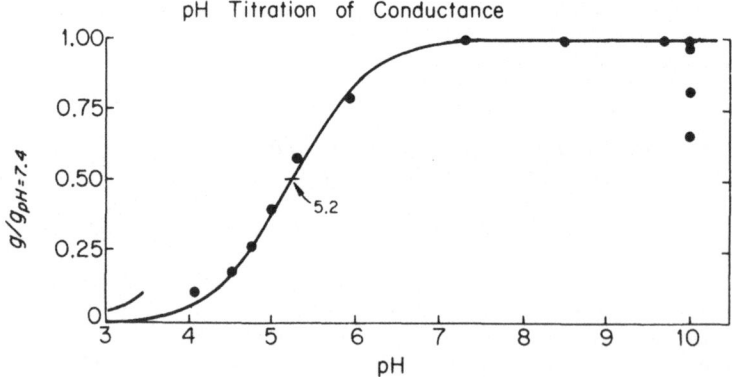

Figure 1. Block of Na channels at low pH. The circles show the relative peak sodium conductance measured in depolarizations to between +20 and +75 mV in a single myelinated nerve fiber under voltage clamp. The external pH was changed from the normal value (7.4) for brief periods using Ringer's solutions buffered with 14 mM glycylglycine–piperazine buffer. The smooth line is the theoretical titration curve of a single acid group with a pK_a of 5.2. (After Hille (16).)

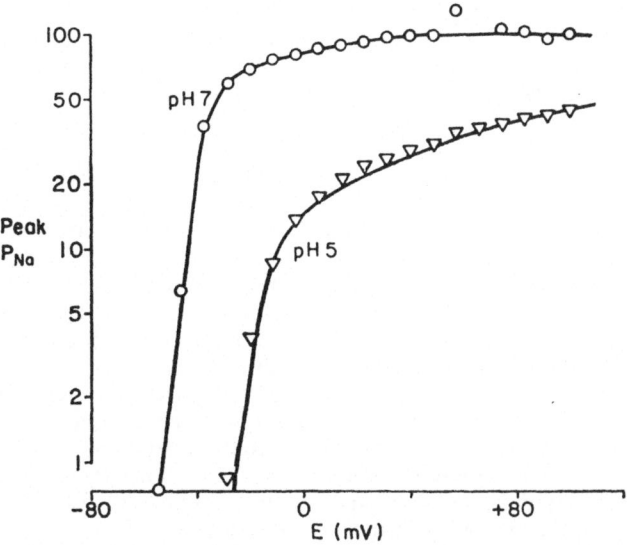

Figure 2. Voltage dependence of Na channel block by low pH. Symbols are the relative peak P_{Na} for various test potentials in a single myelinated nerve fiber under voltage clamp. At pH 7 channels are not blocked by protons and any depolarization beyond 0 mV suffices to open the channels maximally. At pH 5 many channels are blocked and the number open continues to increase with increasing depolarization even beyond +80 mV. The smooth line for pH 5 is derived from the line for pH 7 assuming a 23 mV shift and block depending on voltage as in the quantitative theory of Woodhull (30). (From Woodhull (30).)

partway across the membrane so the proton must cross about 25% of the total transmembrane potential difference to reach the site (30). The site could be partway across the Na channel, although other explanations are possible. The experimental observations have been confirmed by Wagner and Ulbricht (28). Woodhull also found a small voltage-dependent block of Na channels by Ca^{2+} ions that could be explained by a similar mechanism.

PERMEANT IONS INTERACT WITH A NEGATIVE CHARGE IN Na CHANNELS

Selectivity

Chandler and Meves (4) using the squid giant axon were the first to measure the complete alkali ion selec-tivity sequence for Na channels $Li^+ \approx Na^+ > K^+ > Rb^+ > Cs^+$. Following the ideas of Eisenman (7), they suggested that the channel might select for ions of small crystal radius by providing negatively charged sites of high negative field strength, i.e., with small effective radius and a high concentration of charge. The same selectivity sequence for alkali metal ions is observed in myelinated fibers (19) and in frog skeletal muscle membrane (3), and again the idea of a high negative field strength site is the most plausible explanation. One of the simplest examples of such a "site" is the acetate ion in aqueous solution. It has a pK_a of 4.8 and an affinity sequence for metals identical to the permeability sequence of Na channels. Of course the Na channel discriminates among metals far better than the acetate ion does. I believe that this discrimination reflects the more complete dehydration of cations in Na channels than when forming ion pairs with acetate in water.

Energy barrier

Several lines of evidence show that Na^+ ions pass readily through open Na channels without having to cross high energy barriers. The temperature coefficient of fluxes in the channel is almost as low as that for aqueous diffusion (10, 22, 24), and the fluxes in single channels are extraordinarily high (17). At the same time experiments with a variety of organic Na^+-substitutes in myelinated nerve led to the further conclusion that the selectivity filter of the Na channel is an oxygen-lined constriction with a bore of roughly only 3×5 Å (18). In an orifice this small, the permeant alkali metal ions would be largely dehydrated and therefore in a very high free energy state unless otherwise stabilized by strong interactions with the channel. On this basis it was again necessary to postulate a negative charge within the conducting pore of the Na channel, located close to or in the narrow selectivity filter. I consider this to be the same charge that accounts for the alkali metal selectivity sequence. As pointed out by Bezanilla and Armstrong (1) and Hille (19) and elaborated in detail by Hille (20), the interactions determining ionic selectivity in an Na channel must occur at the peak of a rate limiting energy barrier, so the partly dehydrated Na^+ ion passing through the filter and interacting with the negative charge is an "activated complex."

Energy well

Finally permeability and selectivity experiments give evidence for an ion binding site near the outer mouth of the Na channel. The ionic currents carried by some permeant ions, including Li^+, guanidinium, and Tl^+, are smaller than would be predicted from the Goldman-Hodgkin-Katz flux equation, given their permeability relative to Na^+ ions calculated from measured reversal potentials (18, 19). These same ions, and also some impermeant ones including methylguanidinium, reduce the size of inward Na^+ currents when added to an Na-containing Ringer's solution (20). Even when extra Na^+ ions are added to Na-Ringer's the currents are not increased as much as the flux equation predicts. All of these effects are deviations from the "independence relation" of Hodgkin and Huxley (21). The chance that one ion crosses the membrane is being reduced by the presence of other ions. The order of decreasing effectiveness for reducing the current carried by other ions is: Tl^+, guanidinium and relatives, Li^+, Na^+, tetramethylammonium. As with block by protons, the "block" with these other external ions is relieved by large depolarizations.

The selectivity of the channel and deviations from independence may be fitted by a simple energy barrier model shown diagrammatically on the left side of Fig. 3. Here the Na channel is represented as a series of four energy barriers that an ion crosses driven by thermal and electrical forces. A special stipulation is made however that only one ion may be in the channel at a time. Consequently as in enzyme kinetics, the model has saturation kinetics and ion competition. Flux equations may be written down from the barrier model using conventional Eyring rate theory (8), but taking into account saturation. These equations play the same role for this hypothetical system as the Goldman-Hodgkin-Katz equations play for idealized homogeneous constant-field membranes (20).

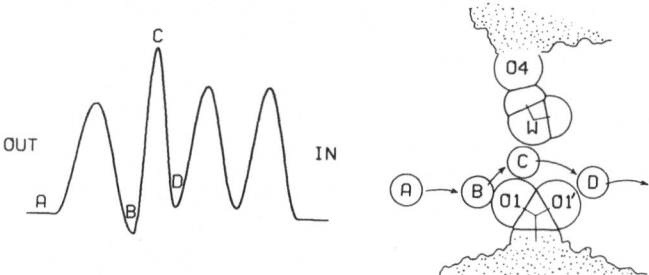

Figure 3. An energy barrier model and molecular interpretation for explaining selectivity and deviations from independence in Na channels. *Left*: Four steps for passing through Na channels are represented as four energy barriers to cross. Binding to site B gives deviations from independence and variation in the peak of the barrier at C gives ionic selectivity. *Right*: Part of the barrier model viewed as partial dehydration and association with the negatively charged carboxylic acid (O1, O1'). A water molecule is shown hydrogen bonded to oxygen O4. Numbering of the oxygens in the selectivity filter (O1, O1', O4) as in Hille (19) and in Fig. 4.

In the barrier model, each ion has its own energy profile The relative heights of barriers in different profiles determine the permeability ratios measured by reversal potentials with mixtures of ions. The depths of wells in individual profiles determine how long each ion lingers in the channel before passing through, or looked at another way, how likely an ion is to block or tie up the channel for a while (20). Remarkably, the observed voltage dependence of the block with different ions is well fitted by choosing a single major binding site (the deepest energy well, labeled B) about 25% of the way across the membrane potential drop, as for protons, and by placing the selectivity filter (the highest barrier, labeled C) just afterwards. For example, the apparent dissociation constant for Tl^+ ions from site B at the resting potential is 22 mM. At 0 mV the value rises to 50 mM. The fitted curves will be presented in a more detailed publication.

Molecular interpretation

The barrier model gives rise to the molecular interpretation drawn on the right side of Fig. 3. The selectivity filter with ionized — COO^- is drawn in side view with an ion moving through positions A, B, C, and D, corresponding schematically to the points labeled on the energy barrier diagram. The aqueous ion (A) loses a small number of water molecules to coordinate (B) with the negative charge at O1. Even methylated cations like methylguanidine can get this far into the channel. In order to permeate, the ion must then move into the short narrow selectivity filter (C) by shedding several more water molecules but still remaining coordinated with the "catalytic" acid group (O1 and O1'). This is the rate limiting "activated complex." Finally the ion emerges into a wider inner end of the channel (D, etc.), and the major hurdles to entering the axon have been crossed. In this view, the ion effectively "crosses the membrane" by moving only a few Ångströms.

TTX AND STX BIND TO A NEGATIVE CHARGE

Hypothesis

In 1965 Kao and Nishiyama (23) proposed that tetrodotoxin blocks by inserting its positively charged guanidinium group into the channel. Once a "molecular" model of the selectivity filter with its negative charge had been suggested (18), it was natural to see if the two hypotheses could be combined. In terms of the diagram of Fig. 3, the TTX molecule might lodge in a position between B and C. This idea works well with the model selectivity filter, since numerous — OH and — NH_2 hydrogens on the toxin line up to form hydrogen bonds to the proposed ring of oxygens of the channel. A drawing of the proposed interaction, first presented at the XXV International Congress of Physiology in Munich in 1971, is shown in Fig. 4. To the right is a space filling model of TTX sitting on the six oxygens of the filter and to the left is the chemical structure of TTX drawn in the same perspective. There are five hydrogen bonds shown in dashes and one electrostatic bond between the toxin (guanidinium) and the negative charge of the filter. A space filling model of STX (29) also fits with one guanidinium group in the filter and the second hovering over it. The fit seems less perfect than with TTX. Three hydrogen bonds and two electrostatic bonds are formed. While the bonds suggested in Fig. 4 may go a long way toward explaining the strong affinity of the receptor for the toxins, the walls of the channel may form additional hydrogen bonds to the "sides" of the toxins as well. The slowness of the reaction of TTX and STX with the receptor (25) might reflect the difficulty of maneuvering the toxin some distance into the channel and achieving the alignment needed for binding. The cases of Na channels resistant to drugs (26) might be explainable without altering the filter by introducing some bulky group in the channel that reaches into the space occupied by the "back" end of the bound toxins.

Experiment

Recently the kinetics and equilibrium binding of TTX and STX have come under intensive study and some properties of the receptor can be deduced. The most important studies are those with binding of radioactivity labeled toxins on unmyelinated nerve bundles in the laboratory of J. M. Ritchie at Yale and those with channel block measured electrophysiologically on single myelinated nerve fibers in the laboratory of W. Ulbricht at Kiel.

First it is clear from binding competition experiments that the TTX receptor is identical to the STX receptor and differs from the local anesthetic receptor (5, 11). At low pH, radioactive toxins bind less well (11, 12, 14). The binding site behaves as an acid with a pK_a of roughly 5.5 that must be ionized for binding to occur. In electrophysiological experiments the block of channels by TTX is also opposed by low pH (27, 28). The apparent pK_a of this effect is voltage dependent with a value near 5.7 at 0 mV. The more positive the potential, the less effective is a given pH at reducing TTX binding. When analyzed by the theory used by Woodhull (30) for proton block of Na channels, the voltage dependence of proton block of TTX binding fits with a proton traversing 38% of the total transmembrane potential difference (28).

Binding of STX and TTX to whole nerve is also opposed by other cations. In order of decreasing inhibitory effectiveness some of the ions are: trivalent ions, Tl^+, Ca^{2+}, Li^+, Na^+, choline (12, 13). The apparent dissociation constant for inhibition by Tl^+ ions is 16–21 mM. Considering the wide variety of tissues involved, these inhibitory effects of metal ions and protons on toxin binding parallel the simultaneous block of channels by metals and protons extraordinarily

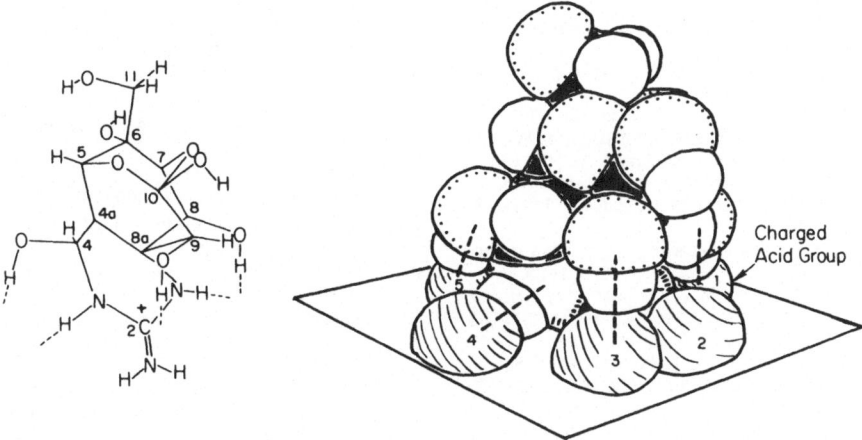

Figure 4. A perspective view of the molecular structure of tetrodotoxin and a space-filling model showing the hypothesis for association with oxygens of the selectivity filter. The two drawings have exactly the same orientation and "magnification." Oxygens of the filter are numbered (1–6) as in Hille (19) and shaded with curved lines. TTX model: Carbons black, oxygens dotted, nitrogens hash-lined, hydrogens all white, hydrogen bonds dashed. The drawing is not meant to imply that oxygens of the selectivity filter are the only components of the TTX receptor.

closely. Evidently the toxin binding site and the selectivity filter are part of the same structure.

CONCLUSION

In conclusion there is much new evidence for a single essential negatively charged group within the Na channel. From its pK_a the group is probably a carboxylic acid. The group coordinates with various metal cations, organic cations, protons, and cationic toxins. The binding seems to be competitive as if only one ion may be there at a time. Numerous impermeant cations can reach the site from the outside, but permeant cations must also continue past this part of the channel to cross the membrane. The affinity of the site for externally applied ligands decreases with increasing depolarization as if part of the transmembrane potential drop occurs in the solution external to the

site. In the neighborhood of the group there probably are numerous other oxygen dipoles. They accept hydrogen bonds from toxins and other organic cations and substitute for shed water molecules to stabilize permeating cations electrostatically. The charge may be thought of as a catalyst for promoting the permeation of acceptable cations through a narrow, and therefore selective, pore. This group is the only chemical group so far "identified" with the Na channel.

REFERENCES

1. BEZANILLA, F., AND C. M. ARMSTRONG. *J. Gen. Physiol.* 60: 588, 1972.
2. CAMOUGIS, G., B. H. TAKMAN AND J. R. P. TASSE. *Science* 156: 1625, 1967.
3. CAMPBELL, D. T. Ionic selectivity of the sodium channel of frog muscle. (Ph.D. Thesis) Seattle: Univ. of Washington, 1974.
4. CHANDLER, W. K., AND H. MEVES. *J. Physiol., London* 180: 788, 1965.

5. COLQUHOUN, D., R. HENDERSON AND J. M. RITCHIE. *J. Physiol., London* 227: 95, 1972.
6. DROUIN, H., AND R. THE. *Arch. Ges. Physiol.* 313: 80, 1969.
7. EISENMAN, G. *Biophys. J.* 2: 259, 1962.
8. EYRING, H., R. LUMRY AND J. W. WOODBURY. *Record Chem. Progr.* 10: 100, 1949.
9. FRANKENHAEUSER, B., AND A. L. HODGKIN. *J. Physiol., London* 137: 218, 1957.
10. FRANKENHAEUSER, B., AND L. E. MOORE. *J. Physiol., London* 169: 438, 1963.
11. HENDERSON, R., J. M. RITCHIE AND G. STRICHARTZ. *J. Physiol., London* 235: 783, 1973.
12. HENDERSON, R., J. M. RITCHIE AND G. STRICHARTZ. *Proc. Natl. Acad. Sci. U.S.* In press.
13. HENDERSON, R., AND G. STRICHARTZ. *J. Physiol. London* 238: 329, 1974.
14. HENDERSON, R., AND J. H. WANG. *Biochemistry* 11: 4565, 1972.
15. HILLE, B. *J. Gen. Physiol.* 51: 199, 1968.
16. HILLE, B. *J. Gen. Physiol.* 51: 221, 1968.
17. HILLE, B. *Progr. Biophys. Mol. Biol.* 21: 1, 1970.
18. HILLE, B. *J. Gen. Physiol.* 58: 599, 1971.
19. HILLE, B. *J. Gen. Physiol.* 59: 637, 1972.
20. HILLE, B. Ionic selectivity of Na and K channels of nerve membranes. Chapt. 4 in: *Membranes—A Series of Advances*, Volume 3, Dynamic Properties of Lipid Bilayers and Biological Membranes, edited by G. Eisenman. New York: Dekker, 1975.
21. HODGKIN, A. L., AND A. F. HUXLEY. *J. Physiol., London* 116: 449, 1952.
22. HODGKIN, A. L., AND A. F. HUXLEY. *J. Physiol., London* 117: 500, 1952.
23. KAO, C. Y., AND A. NISHIYAMA. *J. Physiol., London* 180: 50, 1965.
24. SCHAUF, C. L. *J. Physiol., London* 235: 197, 1973.
25. SCHWARZ, J. R., W. ULBRICHT AND H.-H. WAGNER. *J. Physiol., London* 233: 167, 1973.
26. TWITTY, V. C. *J. Exptl. Zool.* 76: 67, 1937.
27. WAGNER, H.-H., AND W. ULBRICHT. *Arch. Ges. Physiol.* 339: R70, 1973.
28. WAGNER, H.-H., AND W. ULBRICHT. *Arch. Ges. Physiol.* 347: R34, 1974.
29. WONG, J. L., R. OESTERLIN AND H. RAPOPORT. *J. Am. Chem. Soc.* 93: 7344, 1971.
30. WOODHULL, A. M. *J. Gen. Physiol.* 61: 687, 1973.
31. WOODHULL, A. M., AND B. HILLE. Biophys. Soc. Abstracts. 14th Annual Meeting, p. 111a, 1970.

Solutions of the Hodgkin-Huxley equations modified for potassium accumulation in a periaxonal space

WILLIAM J. ADELMAN, JR. AND RICHARD FITZHUGH

Laboratory of Biophysics
National Institute of Neurological Diseases and Stroke
National Institutes of Health, Bethesda, Maryland 20014

ABSTRACT

Hodgkin and Huxley equations were modified to include the properties of an external diffusion barrier separated from the axolemma by a thin periaxonal space in which potassium ions accumulate as a function of membrane activity. Further modifications in the equations took into account new values for $\bar{g}_K$ and new functions for α_n, β_n, α_h, and β_h derived from voltage clamp experiments on *Loligo pealei* giant axons. Equations were solved on a PDP-11 computer using the Gear predictor–corrector numerical method. In comparison with the original Hodgkin and Huxley equations, the modified equations for membrane potentials gave: *1*) more accurate representations of the falling and undershoot phases of the membrane action potential, *2*) more accurate representation of thresholds and latencies, *3*) increases in the periaxonal space potassium ion concentration, K_s, of about 1 mM/impulse, *4*) proper predictions of the time course and magnitude of either undershoot decline or periaxonal potassium ion accumulation during trains of membrane action potentials elicited by repetitive short duration stimuli, and *5*) a somewhat more accurate representation of adaptation (finite train and nonrepetitive responses) during long duration constant current stimulation.—ADELMAN, W. J., JR., AND R. FITZHUGH. Solutions of the Hodgkin-Huxley equations modified for potassium accumulation in a periaxonal space. *Federation Proc.* 34: 1322–1329, 1975.

The Hodgkin and Huxley (14) model for the electrical behavior of the squid giant axon membrane is based on an analysis of an extensive series of voltage clamp experiments. This model is expressed by a set of ordinary differential equations. A number of authors have shown that these equations can be readily solved to produce membrane action potentials for the case of a space clamped axon in which membrane potentials and currents as well as other variables of state of the membrane vary with time but not distance along the axon (4, 5, 8, 14).

In this report, the physical assumptions behind the Hodgkin and Huxley equations (6, 14) have been modified to incorporate an additional variable of state, K_s. K_s is the concentration of potassium ions in the periaxonal space between the outside of the excitable membrane or axolemma and the inside of the Schwann cell sheath, shown to be significant by Frankenhaeuser and Hodgkin (9). K_s continually changes (2) as a result of both potassium current flow through the membrane and diffusion across the Schwann cell layer. The potassium potential across the membrane, E_K, which contributes to the driving force for the membrane potassium current, varies with time as a function of K_s.

In a previous paper (2), it was shown that if E_K varies as a function of K_s, a different value of $\bar{g}_K$, and different functions α_n and β_n, are obtained upon analysis of voltage clamped membrane currents, than those obtained if E_K is constant. The effects of these changes are considered. The rate of change of sodium inactivation, h, has been shown (1) to be a function of external potassium ion concentration as well as membrane potential. The effect of this dependency will also be considered in this work.

In order to compare the effects of each of these modifications to the Hodgkin and Huxley equations, membrane action potentials were calculated using the Hodgkin and Huxley equations as follows: *1*) unmodified; *2*) modified only for K_s; *3*) modified for K_s, new values of $\bar{g}_K$, new functions for α_n and β_n; and *4*) modi-

fied for K_s, new values of $\bar{g}_K$, new functions for α_n, β_n, α_h, and β_h.

Single membrane action potentials in response to both short duration and instantaneous current pulses, trains of action potentials in response to a sequence of instantaneous current pulses at a variety of frequencies (9), and responses to long duration constant current pulses were calculated.

LIST OF SYMBOLS

I = membrane current density, positive outward ($\mu A/cm^2$)

E = membrane potential difference, inside relative to outside (mV).

m = sodium activation (dimensionless, varying between 0 and 1)[1]

h = sodium inactivation (dimensionless, varying between 0 and 1)[2]

n = potassium activation (dimensionless, varying between 0 and 1)[1]

t = time (msec)

C = membrane capacity = 1 $\mu F/cm^2$

ϕ = $3^{(T-6.3)/10}$

T = temperature (°C)

$\bar{g}_{Na}, \bar{g}_K, \bar{g}_L$ = maximal ionic conductances through sodium, potassium and leakage channels, respectively (all in mmho/cm²)

g_m = total membrane conductance = $\bar{g}_{Na}m^3h + \bar{g}_K n^4 + \bar{g}_L$

E_{Na}, E_K, E_L = reversal potentials for sodium, potassium, and leakage current components (all in mV)

K_s = potassium concentration in the phenomenological periaxonal space (mM)

K_0 = potassium concentration in the external bulk solution (mM)

K_i = potassium concentration in the axoplasm (mM)

---

[1] Activation is maximum when m or n = 1.

[2] Inactivation is maximum when h = 0.

E_R = resting membrane potential (mV)

θ = radial thickness of the phenomenological periaxonal space (Å)

$P_K{}^s$ = permeability to potassium of the diffusion barrier between the periaxonal space and the external bulk solution (cm/sec)

EQUATIONS USED

Unmodified equations

The Hodgkin-Huxley equations (5,14) are:

$$\dot{E} = [I - \bar{g}_{Na}m^3h \cdot (E - E_{Na}) - \bar{g}_K n^4 \cdot (E - E_K) - \bar{g}_L \cdot (E - E_L)]/C, \qquad (1)$$

$$\dot{m} = \phi[(1 - m) \cdot \alpha_m(E) - m \cdot \beta_m(E)], \qquad (2)$$

$$\dot{n} = \phi[(1 - h) \cdot \alpha_h(E) - h \cdot \beta_h(E)], \qquad (3)$$

$$\dot{n} = \phi[(1 - n) \cdot \alpha_n(E) - n \cdot \beta_n(E)]. \qquad (4)$$

The relation between the present potential variable E and Hodgkin and Huxley's V is: $V = -60 \text{ mV} - E$. For the unmodified case:

$$\alpha_m(E) = 0.1(-35 - E) \times \left[\exp\left(\frac{-35 - E}{10} \right) - 1 \right]^{-1}, \qquad (5)$$

$$\beta_m(E) = 4 \cdot \exp[(-60 - E)/18], \qquad (6)$$

$$\alpha_h(E) = 0.07 \cdot \exp[(-60 - E)/20], \qquad (7)$$

$$\beta_h(E) = \left[\exp\left(\frac{-30 - E}{10} \right) + 1 \right]^{-1}, \qquad (8)$$

$$\alpha_n(E) = 0.01(-50 - E) \times \left[\exp\left(\frac{-50 - E}{10} \right) - 1 \right]^{-1}, \qquad (9)$$

$$\beta_n(E) = 0.125 \cdot \exp[(-60 - E)/80], \qquad (10)$$

and $\bar{g}_K$ = maximal chord conductance = $I_{K\,max}/(E - E_K)$, assuming E_K a constant.

Modified equations [3]

In all three of the modifications made in this work (cf. 2, 9)

$$E_K = \frac{R(T + 273.15°)}{F} \ln (K_s/K_1), \qquad (11)$$

where

R = 8.31434 joule (mole°)$^{-1}$,

F = 9.64867·10^4 coulomb/mole, and

$$\frac{dK_s}{dt} = (1/\theta)[(I_K/F) - P_K{}^s(K_s - K_o)]. \qquad (12)$$

For dimensional consistency in these equations, the numerical value of θ used in computation must be 100 times its value in Å, that for $P_K{}^s$ must be 10^{-3} times its value in cm/sec, and that for F must be 96.4867.

Adelman, Palti and Senft (2; *equations 19, 20*), assuming the present *equation 11*, analyzed the experiment shown in their Fig. 8, to give a higher value of $\bar{g}_K$ than given by assuming a constant E_K. On this basis, they also redefined α_n and β_n (here corrected from 4.5 C to 6.3 C):

$$\alpha_n(E) = (0.005606)[7.93 - (E + 60)] \Big/ \left[\exp\left(\frac{7.93 - (E + 60)}{7.93} \right) - 1 \right], \qquad (13)$$

and

$$\beta_n(E) = (0.121866) \times \exp[(-E - 60)/26.72]. \qquad (14)$$

Adelman and Palti (1) redefined α_h and β_h:

---

[3] Potential differences across the external barrier have been omitted. The resistance of the barrier makes the recorded potential differ from the membrane potential E by only a small component proportional to the current. Following a stimulus pulse, this is zero, while during constant current stimulation, a barrier resistance of 1.8 ohm cm^2 would shift the curves of Fig. 10E and 11F upward by only 0.1 mV. Any emf arising from differences of concentration of sodium and potassium across the barrier have been neglected.

$$\alpha_h(E, K_s) = [0.171385 - 0.0380856 \ln (K_s)]$$

$$\times \exp \left[\frac{-60 - E}{27.4} \right] , \quad (15)$$

which is their equation 7 in natural log form corrected from 3.5 C to 6.3 C, and

$$\beta_h(E, K_s) = 1.3602/[1 + \exp(-4 - 0.11111 \cdot E)]$$

$$+ 0.013602 \exp \left[\frac{-E \cdot K_s}{32.5 \cdot K_s + 185} \right] , \quad (16)$$

which is the combination of their equations 9 and 10 corrected from 3.5 C to 6.3 C. In both *equations 15* and *16*, K_s replaces K_o in the functions defined by Adelman and Palti (1).

METHODS OF COMPUTATION

The differential equations were solved using the numerical method of Gear (10, 11), programmed in FORTRAN on a PDP-11 computer. Gear's method is a self-starting predictor–corrector method with the order of the method and the step size automatically adjusted to achieve the desired accuracy and speed. The alternate method of Gear for stiff equations (those with widely differing relaxation times in different variables) was not used. The accuracy achieved cannot be stated precisely, but amounted to roughly five significant figures.

In these computations, for the most part, values of significant constants (maximal conductances, reversal potentials, and so on) were taken from voltage clamp data obtained from giant axons of *Loligo pealei* (2). Membrane action potentials recorded from these axons were compared with computed membrane action potentials whenever possible. Computed results were taken from core memory of the PDP-11 through a D/A converter and plotted on an x/y plotter by means

of plot routine devised by our colleague, Dr. Brad J. Cox.

MEMBRANE ACTION POTENTIALS

A typical membrane action potential recorded from a giant axon of *Loligo pealei* is shown in the upper part of Fig. 1. This just above threshold response has a long latency, a smooth rise to peak, and a somewhat linear fall. The falling phase gently rounds into the undershoot phase, the negative value of which is -70 mV. This response was recorded at 4.5 C by Adelman, Palti and Senft from axon 3 (see Tables II and III in ref. 2). This axon had a resting potential, E_r, of -60 mV. Analysis of membrane currents obtained upon voltage clamping this axon indicated that $E_{Na} = 55$ mV, $\bar{g}_{Na} = 85.5$ mmho/cm^2, $\bar{g}_K = 42$ mmho/cm^2, and $\bar{g}_L = 0.7$ mmho/cm^2. Table III in Adelman, Palti and Senft (2) gives the value of the thickness of the Frankenhaeuser and Hodgkin space (9), θ, as 379 Å, and the values of $P_K{}^s$, K_o, K_i as 3.41×10^{-4} cm/sec, 10 mM and 284 mM, respectively, for axon 3.

A reconstruction of the membrane action potential of axon 3 is shown in the middle part of Fig. 1. This reconstruction was obtained using the Hodgkin and Huxley (14) *equations 1 through 4*, modified for K_s, new values of $\bar{g}_K$, and new functions for α_n, β_n, α_h, and β_h, as given in *equations 11 through 16*. Henceforth in this paper, these equations will be referred to as the fully modified Hodgkin and Huxley equations. In the calculation, the values of E_{Na}, $\bar{g}_{Na}$, $\bar{g}_L$, E_r, K_o, K_i, θ, $P_K{}^s$, and T were the experimentally determined values listed above for axon 3. The stimulating current was 28.8 μamp/cm^2 flowing for 0.45 msec.

Upon comparing the reconstructed response (Fig. 1, middle) with the

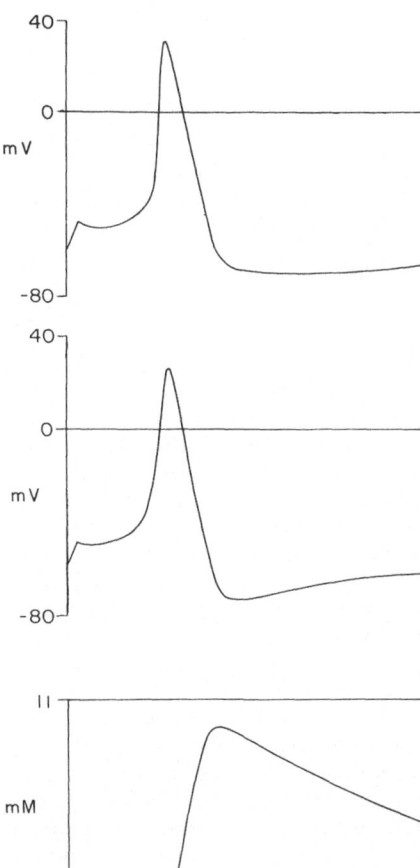

Figure 1. *Upper*: membrane action potential recorded from a *Loligo pealei* giant axon. *Middle*: membrane action potential computed with Hodgkin and Huxley equations modified for K_s, a new value of $\bar{g}_K$, and new functions for α_n, β_n, α_h, and β_h using constants derived from an analysis of membrane currents obtained upon voltage clamping the axon whose action potential is shown in the upper record. *Lower*: values of the periaxonal space potassium ion concentration (ordinate) computed coincident with the action potential shown in the middle record. The time base on the abscissa of the lower record is appropriate to all three records. See text for further details.

recorded response (Fig. 1, top), a number of similarities are noticed. The threshold, latency, rate of rise, peak, and falling phase of the reconstructed response are quite similar to those in the recorded response. However, the undershoots are dissimilar. The calculated membrane action potential has a more negative (by about 5 mV) and less persistent undershoot.

Computed values of the concentration of potassium ions in the peri-axonal space, K_s, are plotted in the lower part of Fig. 1. According to this calculation, a single action potential increases K_s from just above 10 mM at rest to almost 11 mM at the peak of the potassium accumulation phenomenon. The decline from the peak value of K_s has a half-time of 10 msec. If the decline of potassium concentration in the space were governed only by diffusion through the barrier, the time constant τ_K of its decline following a single action potential would be given by the relation (9):

$$\tau_K = \theta/P_K{}^s. \qquad (17)$$

For axon 3, $\theta = 379$ Å and $P_K{}^s = 3.41 \times 10^{-4}$ cm/sec. Therefore, on this basis, $\tau_K = 11.1$ msec.

If we assume that the decline of K_s shown in Fig. 1 (lower) is singly exponential, we can calculate τ_K using the following relation (9):

$$\tau_K = t/\ln\,(y_0/y), \qquad (18)$$

where $y_0 = \Delta K_s\,(0)$, and $y = \Delta K_s\,(t)$. ΔK_s is the excess concentration of potassium ($K_s - K_o$) in mM. A semilogarithmic plot of the decline of ΔK_s versus t was made and τ_K was calculated using *equation 18* over the range of values corresponding to the linear portion of this plot. On this basis, $\tau_K = 13.2$ msec. This is larger than the value just calculated because potassium is flowing into the space through the nerve membrane, as well as diffusing out through the barrier.

The computing routine for reconstructing the action potential also gives values of Hodgkin and Huxley parameters m, h, and n as a function of time. Parameter n reaches a maximum value towards the end of the falling phase of the action potential, declines during the beginning of the undershoot with a half-time of about 5 msec, and then declines more slowly throughout the remainder of the undershoot phase. It is obvious that during most of the late spike and early undershoot phases of the action potential, the values of K_s plotted in Fig. 1 are determined by the resultant of membrane potassium currents and potassium ion diffusion across the outer diffusion barrier. Therefore, single exponential decline in K_s values should not be reached until the membrane potassium current becomes small.

In Fig. 2 a membrane action potential computed with the unmodified Hodgkin and Huxley equations (A) is compared with an action potential computed with the fully modified Hodgkin and Huxley equations (B). In both of these computations, the membrane potential was stepped from the resting potential of -60 mV to -49.5 mV by the application of an instantaneous current pulse. The computation then proceeded in the normal manner. Values of $\bar{g}_{Na}$, $\bar{g}_L$, E_{Na}, and T were 120 mmho/cm², 0.24 mmho/cm², 39 mV, and 4.5 C, respectively, in both computations. In computing the response shown in A, $\bar{g}_K = 40$ mmho/cm² and $E_K = -80$ mV. In computing the response shown in B, $\bar{g}_K = 62.5$ mmho/cm², $K_o = 10$ mM, $K_i = 258$ mM, $\theta = 379$ Å, and $P_K{}^s = 2.41 \times 10^{-4}$ cm/sec. These values were selected to give close correspondence between the rising phases and peaks of the action potentials computed with both sets of equations.

Examination of Fig. 2 reveals that the action potential in A has a distinct hump on the falling phase that is much less pronounced on the falling phase of the action potential in B. The most negative value of the membrane potential during the undershoot phase of the action potential in A is -78.4 mV, which is very close to the constant value of E_K (-80 mV) used in the computation. The most negative value of the undershoot phase of the action potential in B is -73.9 mV, almost 5 mV more positive than that in A. The maximal value of the periaxonal potassium concentration, K_s, computed for the action potential in B was 11.2 mM. Action potentials computed with the fully modified Hodgkin and Huxley equations using instantaneous current pulses were almost identical to those computed with these equations using finite short duration current pulses. We will use the instantaneous current pulse stimulation later in this paper for computing the

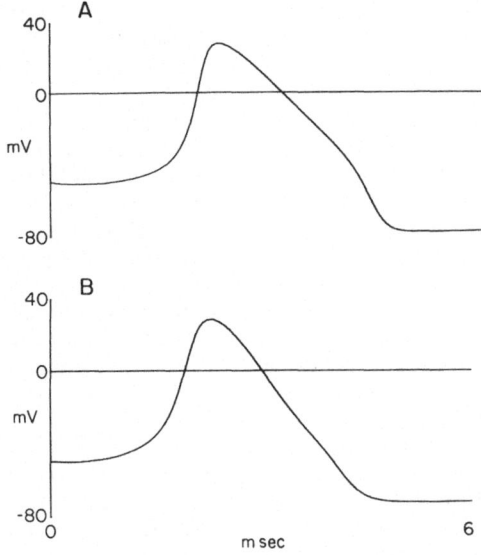

Figure 2. *A*: membrane action potential computed with the unmodified Hodgkin and Huxley equations. *B*: membrane action potential computed with the fully modified Hodgkin and Huxley equations. The time base in *B* is linear and is appropriate to both *A* and *B*. See text for further details.

responses to a train of short repetitive stimuli.

Figure 3 plots membrane action potentials computed with the Hodgkin and Huxley equations modified only for K_s (*A*) and modified for K_s, a new value of $\bar{g}_K$, and new functions for α_n and β_n (*B*). The values of $\bar{g}_{Na}$, $\bar{g}_L$, E_{Na}, and T used in both computations were the same as those used in the computations shown in Fig. 2. In Fig. 3*A*, $\bar{g}_K = 40$ mmho/cm², and in Fig. 3*B*, $\bar{g}_K = 62.5$ mmho/cm². In both *A* and *B*, $K_0 = 10$ mM, $K_i = 258$ mM, $\theta = 379$ Å, and $P_K^s = 2.41 \times 10^{-4}$ cm/sec. The action potential in *A* is rather similar to the Hodgkin and Huxley action potential shown in Fig. 2*A* in having a hump on the falling phase and a rather sharp transition into the undershoot phase. However, the most negative value of the undershoot of the action potential shown in Fig. 3*A* is more positive than that in Fig. 2*A*.

In Fig. 3*B*, the computed membrane action potential has a distinctly prolonged overshoot. This is not characteristic of any membrane action potential recorded in normal artificial sea water (for composition see Table I in ref 2). Of the four computed action potentials illustrated in Figs. 2 and 3, the one calculated with the fully modified Hodgkin and Huxley equations most closely resembles that recorded from giant axons from *Loligo pealei*.

ACTION POTENTIAL TRAINS

Frankenhaeuser and Hodgkin (9) elicited trains of action potentials from giant axons of *Loligo forbesi* by repetitive (50/sec) short duration stimuli. They showed that at the beginning of a train of impulses the undershoot phases of the action potentials became successively less negative and approached a steady state value in an exponential manner with a time constant of 30 to 100 msec. On the basis that the membrane potential during the undershoot phase approaches the potassium potential, they correlated the decline in undershoot negativity with a rise in potassium ion concentration in a phenomenological periaxonal space separated from the bulk of the external solution by an external diffusion barrier.

The unmodified Hodgkin and Huxley equations have no means by which the findings of Frankenhaeuser and Hodgkin (9) can be predicted. Figure 4 is a plot of a train of membrane action potentials computed using the unmodified Hodgkin and Huxley equations for repetitive stimulation with 20 mV instantaneous depolarizations at a frequency of 50/sec. Notice that both the overshoot and undershoot phases have maximal

values that are invariable throughout the train.

Figure 5 illustrates a typical recorded train of membrane action potentials elicited from a *Loligo pealei* giant axon by 0.5 msec currents at a frequency of 50/sec. In contrast to the computed responses in Fig. 4, the undershoot phase becomes less negative with each successive response. If we assume that these values decline exponentially we can determine the time constant, τ, of the decline of maximal negativity of the undershoot phases from the relation:

$$\tau = t/\ln\{(y(0) - y(\text{ss}))/(y - y(\text{ss}))\}, \quad (19)$$

where $y(0)$ is the most negative value of the first spike undershoot, y is the most negative value of any undershoot at time, t, the time elapsed since the first spike undershoot, and $y(\text{ss})$ is the value that y approaches in the steady state. Upon plotting

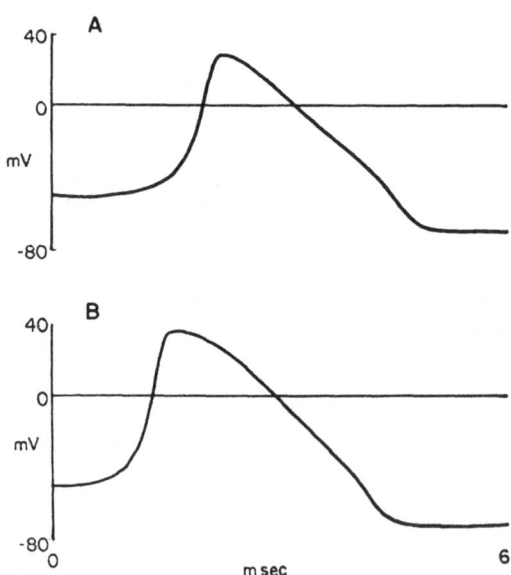

Figure 3. *A*: membrane action potential computed with the Hodgkin and Huxley equations modified for K_s only. *B*: membrane action potential computed with the Hodgkin and Huxley equations modified for K_s, a new value for $\bar{g}_K$, and new functions for α_n, and β_n. See text for further details.

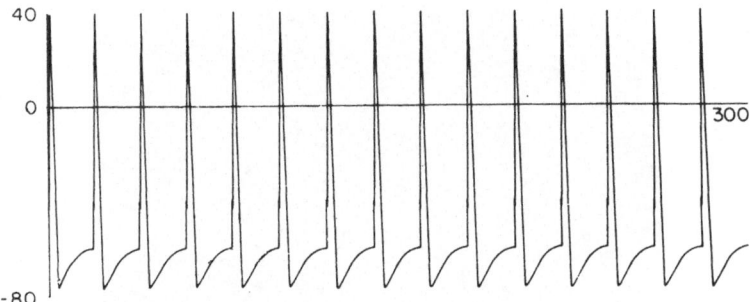

Figure 4. Train of membrane action potentials in response to repetitive stimulation with instantaneous current pulses at 50/sec, computed using the unmodified Hodgkin and Huxley equations. Values of constants were: $\bar{g}_{Na}$ = 120 mmho/cm^2, $\bar{g}_K$ = 24.4 mmho/cm^2, $\bar{g}_L$ = 0.82 mmho/cm^2, E_{Na} = 48.5 mV, E_K = −80.2 mV, E_r = −60 mV and T = 5 C. Computed membrane potential in mV (ordinate) plotted linearly against time (abscissa).

values of $y - y(ss)$ semilogarithmically against time, a linear plot was obtained indicating that *equation 19* was an appropriate description of the decline of undershoot negativity. The fit of *equation 19* to the undershoot data gave a value of 18.8 msec for τ. This value is below the range of values for τ obtained by Frankenhaeuser and Hodgkin (9) for similar spike trains in *Loligo forbesi* axons, and it is only about ⅓ of the mean value obtained by Frankenhaeuser and Hodgkin. This is to be expected from the higher value of P_K^s determined by Adelman et al. (2) for *Loligo pealei* axons than that obtained for P_K^s in *Loligo forbesi* axons by Frankenhaeuser and Hodgkin (9).

Figure 6 illustrates a typical recorded train of membrane action potentials elicited with 100/sec stimuli. Notice the severe decline in action potential amplitudes, the decrease in negativity of successive undershoot values, and the periodic failure of the stimulus to reach threshold. Notice also that after each ineffective stimulus both the action potential amplitude and the undershoot negativity tend to return toward initial values.

Later we will show that this type of behavior can be reconstructed from the fully modified Hodgkin and Huxley equations.

Figure 7A illustrates a reconstruction of a repetitive train of membrane action potentials using the fully modified Hodgkin and Huxley equations. Stimuli were 50/sec instantaneous current steps giving 20 mV depolarizations. The decline in undershoot values in Fig. 7A is very similar to the decline illustrated in Fig. 5. Figure 7B plots the computed values of the potassium concentration, K_s, in the periaxonal space. Notice that the calculated peak values of K_s reach a steady state after five responses. We fit *equation 19* to the undershoot values in Fig. 7C and obtained a value of 15.6 msec for τ. This value is rather similar to the value of 18.8 msec obtained upon fitting *equation 19* to the undershoot data in Fig. 5. To all intents and purposes, the reconstruction of the response train in Fig. 7 from the fully modified Hodgkin and Huxley equations must be considered as a reasonable approximation of the recorded responses shown in Fig. 5,

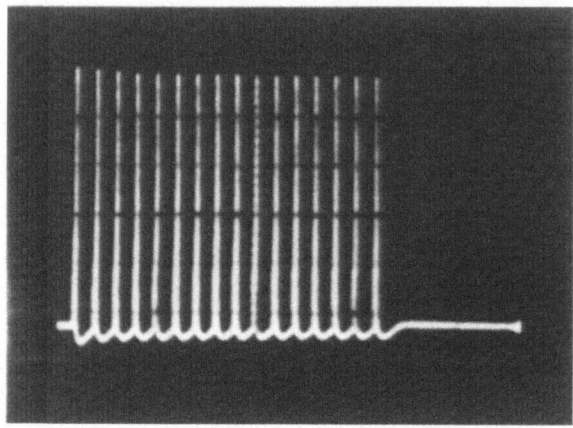

Figure 5. Recorded train of membrane action potentials elicited by repetitive stimulation of a *Loligo pealei* giant axon with a sequence of short depolarizing currents at a frequency of 50/sec. 20 mV and 50 msec per major grid divisions. $T = 8.5$ C and $E_r = -62$ mV.

considering the 3.5 C temperature difference between recorded and computed responses.

Figure 8*A* is a plot of responses to a train of instantaneous currents at a frequency of 100/sec computed using the fully modified Hodgkin and Huxley equations. Notice the decline in overshoot amplitude, the failure of every fourth stimulus to reach threshold, the recovery of spike amplitude following each ineffective stimulus, and the pattern of undershoot values. All these are similar in pattern to those seen in the recorded train shown in Fig. 6. Fig. 8*B* is a

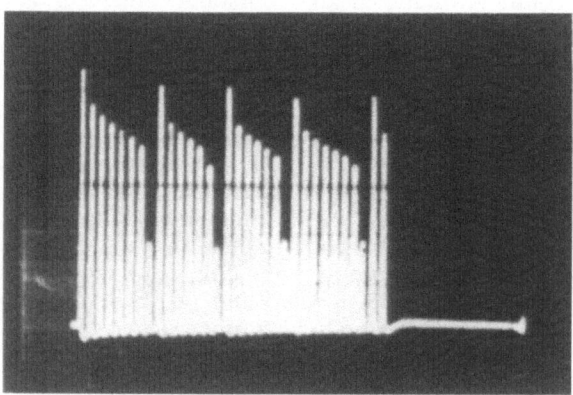

Figure 6. Recorded train of membrane action potentials elicited by repetitive stimulation of a *Loligo pealei* giant axon with a sequence of short depolarizing currents at a frequency of 100/sec. 20 mV and 50 msec per major grid divisions. $T = 8.5$ C and $E_r = -62$ mV.

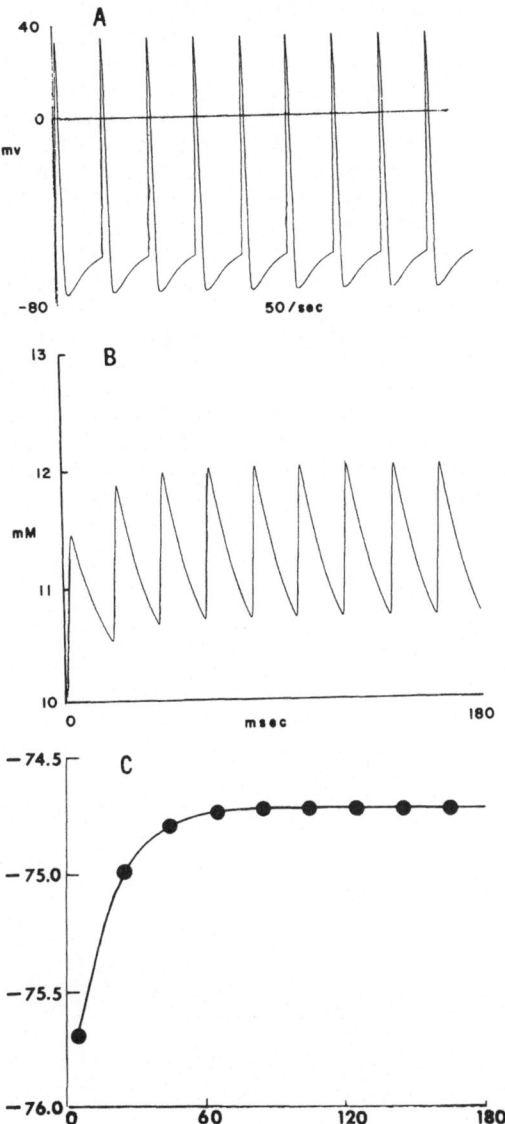

Figure 7. A: train of membrane action potentials in response to instantaneous current pulses at a frequency of 50/sec, computed using the fully modified Hodgkin and Huxley equations. Constants used were: $\bar{g}_{Na} = 152$ mmho/cm^2, $\bar{g}_K = 62.5$ mmho/cm^2, $\bar{g}_L = 0.24$ mmho/cm^2, $E_{Na} = 39$ mV, $E_r = -60$ mV, $K_o = 10$ mM, $K_i = 284$ mM, $\theta = 379$ Å, $P_K^s = 2.41 \times 10^{-4}$ cm/sec, and $T = 5$ C. B: computed values of K_s in mM (ordinate) associated with the computation shown in A plotted linearly against time (abscissa). C: Maximal values of undershoot negativity taken from the computer readout plotted in A. Ordinate in mV; abscissa in msec.

plot of K_s values computed during the response train shown in Fig. 8*A*.

In addition to reconstructing trains of responses recorded from *Loligo pealei* axons we have been able to represent train behavior recorded by Frankenhaeuser and Hodgkin (9) from *Loligo forbesi* axons by using the fully modified Hodgkin and Huxley equations with values of θ and $P_K{}^s$ appropriate for *Loligo forbesi* axons (9). Figure 9 (*upper*) illustrates a train of responses computed using the fully modified Hodgkin and Huxley equations for instantaneous current stimulation at a frequency of 50/sec. In this case, $\theta = 270$ Å and $P_K{}^s = 6 \times 10^{-5}$ cm/sec. The exponential decline in undershoot values has a longer time constant than the time constant for undershoot decline in the train shown in Fig. 7. In Fig. 9 (*lower*) computed values of K_s (continuous line) are plotted throughout the train. The points are values of K_s taken from Fig. 8 in Frankenhaeuser and Hodgkin's paper (9) for a train of responses to 50/sec stimuli. Frankenhaeuser and Hodgkin determined these points by calibrating the maximum negativity of single response undershoot values against a variety of external potassium ion concentrations. The correspondence between our computed values of K_s and those determined by Frankenhaeuser and Hodgkin (6) is quite good.

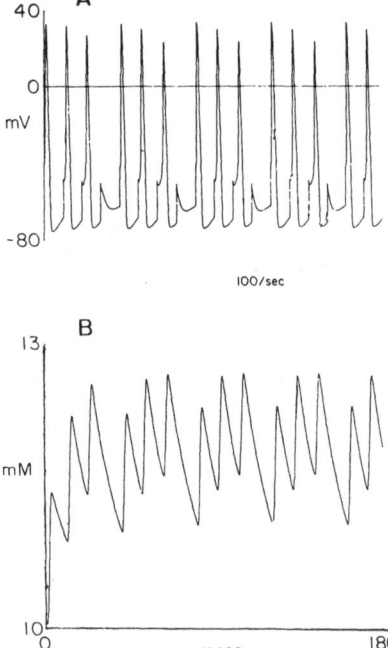

Figure 8. *A*: train of membrane action potentials in response to instantaneous current pulses at a frequency of 100/sec, computed using the fully modified Hodgkin and Huxley equations. Values of constants are the same as in Fig. 7. *B*: computed values of K_s in mM (ordinate) associated with the computation shown in A plotted linearly against time (abscissa).

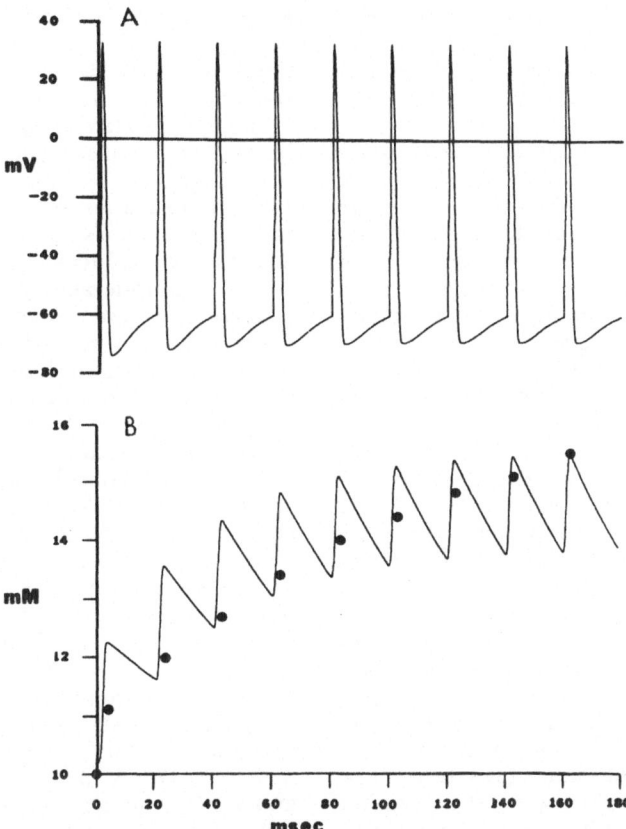

Figure 9. *A*: train of membrane action potentials in response to instantaneous current pulses at a frequency of 50/sec, computed using the fully modified Hodgkin and Huxley equations with values for θ and P_K^s taken from Frankenhaeuser and Hodgkin (9). All other constants are the same as Fig. 7. *B*: computed values of K_s in mM (continuous line) and values (points) of K_s taken from Frankenhaeuser and Hodgkin (9) plotted linearly against time. See text.

RESPONSES TO LONG DURATION CONSTANT CURRENT STIMULATION

Computations using the unmodified Hodgkin and Huxley equations for constant current steps over a wide range of values give infinite trains of impulses (7). Finite trains are only seen in these computations as a borderline phenomenon occurring over a very narrow span of current values, between the current required to produce one impulse and that for infinite trains. Agin (3) has reported that the Hodgkin and Huxley equations predict a range of repetitive firing frequencies that are related to the logarithm of the intensity of the constant current stimulus. Stein (15) confirmed this finding, but reported that the lowest frequency calculated with the Hodgkin and Huxley equations was about 50/sec with a discontinuous change in frequency of discharge from 0 to over 50/sec just above a stimulus intensity of 6.5 μamp/cm^2.

In an extensive study of *Loligo pealei* axons bathed in normal sea water, Hagiwara and Oomura (13) found that only short finite trains of impulses occur in response to constant current stimulation, even for a considerable range of current values. However, Hagiwara and Oomura state that not all squid giant axons fire repetitively. Recently, Hagiwara (personal communication) informed us that repetitive firing had been found to be the exception rather than the rule. Recently, Cox and Adelman (ms in preparation) have found that freshly excised *Loligo pealei* axons only fire repetitively when the external calcium ion concentration is reduced below the normal 10 mM value. Guttman and Barnhill (12) induced repetitive firing in *Loligo pealei* axons bathed in magnesium-free artificial sea water solutions. They found that repetitive firing was related to the state of health of the animals from which the axons were excised. Cox and Adelman (ms in preparation) found that repetitive firing occurred more frequently in axons excised from squid collected in the spring from cold waters than in axons excised from squid collected in the late summer from warmer waters. In no case could they induce repetitive firing in axons bathed in sea water solutions containing "normal" concentrations of calcium (10 mM) and magnesium (50 mM).

In 1961, FitzHugh (7) stated that a likelihood exists that there is some accommodative process slower than any in the original Hodgkin and Huxley equations that accounts for the behavior of the real excised axon. He suggested that a slow process re-

lated to the accumulation of potassium ions in the periaxonal space might be a candidate for this accommodation. As the fully modified Hodgkin and Huxley equations used in this work consider, on the basis of voltage clamp data, the full effects of potassium ion accumulation in the periaxonal space on the membrane ionic conductance parameters, it was of considerable interest to see if these equations would give either nonrepetitive firing to constant current stimulation, or finite train behavior. The effects of potassium accumulation, per se, on repetitive firing were also examined.

This later consideration is examined initially. Figure 10 plots the results of a set of computations for 45 msec constant current stimulation using the Hodgkin and Huxley equations modified only for the additional variable of state, K_s. Values of $\bar{g}_{Na}$, $\bar{g}_K$, $\bar{g}_L$, E_{Na}, E_r, K_o, K_i, θ, $P_K{}^s$, and T used in the computation were 90 mmho/cm^2, 24.4 mmho/cm^2, 0.7 mmho/cm^2, 55 mV, −60 mV, 10 mM, 284 mM, 379 Å, 3.4 × 10^{-4} cm/sec, and 4.5 C, respectively. These values were obtained from axon 3 by Adelman et al. (2). The value of $\bar{g}_K$ was based on the chord potassium conductance data given in Fig. 8B in Adelman et al. (2).

The frequency of the trains, f, at 4.5 C, shown in Fig. 10 *C, D,* and *E*, is fit by the equation:

$$f = 21.4 \ln(I_m + 1), \qquad (20)$$

where I_m is the current intensity in μamp/cm^2. Agin (3) fit the frequency of computed responses to constant current stimulation using the unmodified Hodgkin and Huxley equations at 6.3 C with the following relation:

$$f = 27 \ln(I_m + 1). \qquad (21)$$

If we apply a Q_{10} of 3 to adjust

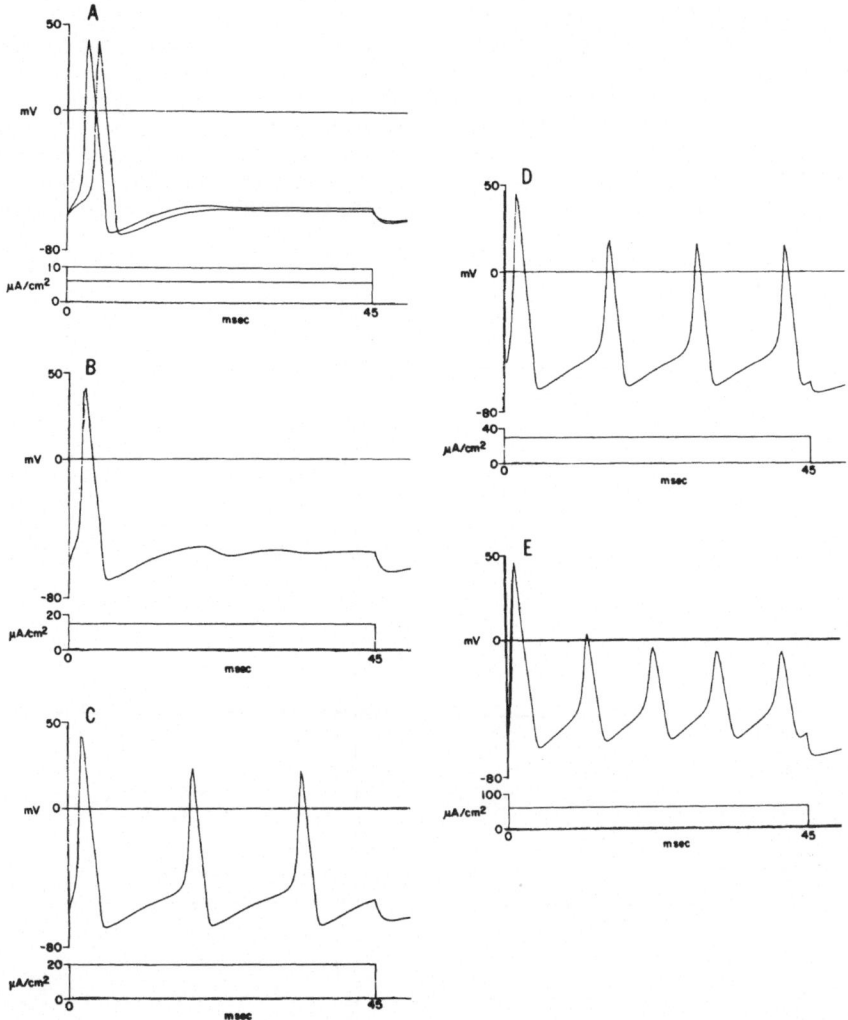

Figure 10. Set of membrane potential computations using the Hodgkin and Huxley equations modified for K_s only. Constant current stimulation. Values of currents are plotted below each membrane potential record. See text for further details.

equation 21 from 6.3 C to 4.5 C, the frequency coefficient is reduced from 27 to 22.2. If we use Guttman and Barnhill's (12) experimentally determined Q_{10} of 2.7 then the coefficient is reduced from 27 to 22.6.

It would seem apparent that the introduction of the new variable of state, K_s, into the Hodgkin and Huxley equations does not alter significantly the repetitive firing frequency from that predicted by the unmodified

Hodgkin and Huxley equations (14).

The maximal K_s values achieved during the trains of repetitive responses illustrated in Fig. 10 D and E are 13 and 14 mM, respectively. It is apparent that these K_s values are not sufficient to significantly influence repetitive firing behavior.

Results of computations using the fully modified Hodgkin and Huxley equations for constant current stimulation are quite different than those shown in Fig. 10. Figure 11 illustrates the results of a set of computations using the fully modified Hodgkin and Huxley equations.

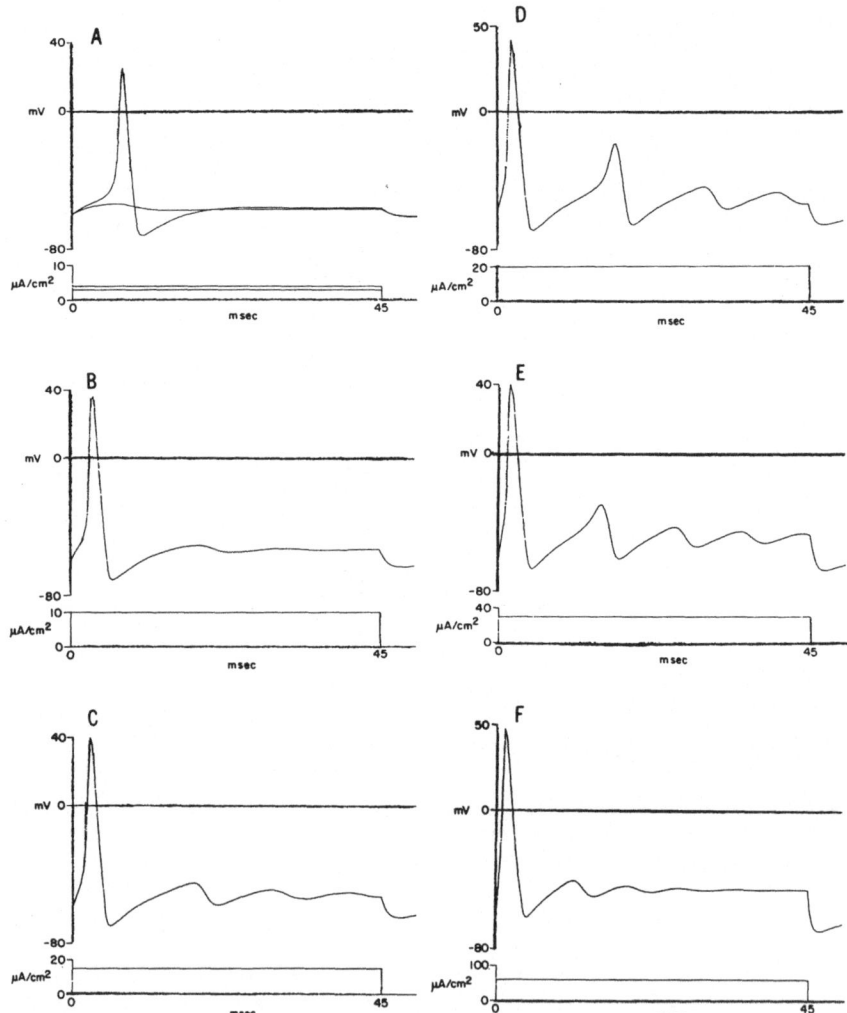

Figure 11. Set of membrane potential computations using the fully modified Hodgkin and Huxley equations. Constant current stimulation. Values of currents are plotted below each membrane potential record. See text for further details.

Stimuli are 45 msec constant currents. Values of $\bar{g}_{Na}$, $\bar{g}_K$, $\bar{g}_L$, E_{Na}, E_r, K_o, K_i, θ, $P_K{}^s$, and T used in the computation were 90 mmho/cm², 42 mmho/cm², 0.7 mmho/cm², 55 mV, −60 mV, 10 mM, 284 mM, 379 Å, 3.4×10^{-4} cm/sec and 4.5 C, respectively. These values were obtained from axon 3 (2). The value for $\bar{g}_K$ was based on Fig. 8A in Adelman et al. (2). Notice that repetitive firing is almost totally suppressed in Fig. 11. Only one additional, rather small, response occurs in D and E.

The computed responses illustrated in Fig. 11 are intermediate between the finite trains recorded by Hagiwara and Oomura (13) and the nonrepetitive behavior seen by Cox and Adelman (ms in preparation). Cox and Adelman found that responses similar to those illustrated in Fig. 11 B and F were the rule for constant current stimulation of *Loligo pealei* axons bathed in normal artificial sea water (cf. Table I, ref 2).

The maximal K_s values produced in this computation were just under 12 mM. As these values are less than those computed for constant current stimulation with the Hodgkin and Huxley equations modified for K_s only, it is apparent that potassium accumulation in the periaxonal space is not the reason, per se, for nonrepetitive firing behavior. Examination of the values of m, h, and n computed for the cases illustrated in Figs. 10 and 11 reveals that the major factor related to adaptation and nonrepetitive firing is parameter h. The Adelman and Palti (1) equations for α_h and β_h (*equations 15* and *16*, this paper) used in the computation illustrated in Fig. 11 are rather different than the original α_h and β_h equations of Hodgkin and Huxley (14). Values of h computed with the fully modified Hodgkin and Huxley equations were usually lower than those computed with the Hodgkin

and Huxley equations modified for K_s only. In addition, values of h decreased more rapidly during the transition from the rising phase of the second response to its falling phase (Fig. 11 D and E) in the fully modified computations than during the same transition (Fig. 10 C, D, and E) computed using only the K_s modification. Therefore, the slow accommodative process postulated by Fitz-Hugh (7) would seem to be related to the rate of change of sodium inactivation as a function of both K_s and membrane potential (1).

SIGNIFICANCE OF THE COMPUTATIONS

In general, the fully modified Hodgkin and Huxley equations can be said to be better predictors of squid axon behavior than the original Hodgkin and Huxley equations. Not only can the membrane action potential be reconstructed for *Loligo pealei* axons with greater fidelity, but a variety of neurophysiological phenomena, such as responses to trains of stimuli and responses to constant current stimulation, can be represented in a more realistic manner.

This paper is dedicated to Dr. Kenneth S. Cole, who pioneered the use of computers in solving the Hodgkin-Huxley equations. We thank Drs. Wilfred Rall and Clifford S. Patlak for reading the manuscript and making suggestions for its improvement.

REFERENCES

1. ADELMAN, W. J., JR., AND Y. PALTI. *J. Gen. Physiol.* 53: 685, 1969.
2. ADELMAN, W. J., JR., Y. PALTI AND J. P. SENFT. *J. Memb. Biol.* 13: 387, 1973.
3. AGIN, D. *Nature* 201: 625, 1964.
4. COLE, K. S., H. A. ANTOSIEWICZ AND P. RABINOWITZ. *J. Soc. Ind. Appl. Math.* 3: 153, 1955.
5. FITZHUGH, R. *J. Gen. Physiol.* 43: 867, 1960.
6. FITZHUGH, R. In: *Biological Engineering*, edited by H. P. Schwan. New York: McGraw-Hill, 1969, Chapt. 1.

7. FITZHUGH, R. *Biophys. J.* 1: 445, 1961.
8. FITZHUGH, R., AND H. A. ANTOSIEWICZ. *J. Soc. Ind. Appl. Math.* 7: 447, 1959.
9. FRANKENHAEUSER, B., AND A. L. HODGKIN. *J. Physiol., London* 131: 341, 1956.
10. GEAR, C. W. *Commun. Assoc. Comput. Mach.* 14: 185, 1971.
11. GEAR, C. W. *Numerical Initial Value Problems in Ordinary Differential Equations.* Engle-

wood Cliffs, N.J.: Prentice-Hall, 1971.
12. GUTTMAN, R., AND R. BARNHILL. *J. Gen. Physiol.* 55: 104, 1970.
13. HAGIWARA, S. AND Y. OOMURA. *Japan. J. Physiol.* 8: 234, 1958.
14. HODGKIN, A. L., AND A. F. HUXLEY. *J. Physiol. London* 117: 500, 1952.
15. STEIN, R. B. *Proc. Roy. Soc. London Ser. B* 167: 64, 1967.

Noise measurements in axon membranes[1]

HARVEY M. FISHMAN

Department of Physiology and Biophysics
University of Texas Medical Branch, Galveston, Texas 77550

ABSTRACT

A fluctuation (noise) component, which arises from spontaneous microscopic conductance perturbations, about the mean conduction characteristic of small areas of axon membrane has been measured and appears to relate to voltage-dependent potassium-ion movements. Since the kinetics of relaxing membrane structures which produce ion permeation noise may be deduced from the form of power density spectra of fluctuations, a comparison of measured and theoretical noise spectra may lead to direct tests of conduction models.—FISHMAN, H. M. Noise measurements in axon membranes. *Federation Proc.* 34: 1330–1337, 1975.

Among the many contributions of "Kacy" Cole to membrane biophysics, two stand out as fundamental. His impedance work not only provided a measure of membrane capacitance and conductance, but also established conduction as the basis for all excitable phenomena. From the near constancy of membrane capacitance and the unlikelihood of uniform membrane conduction without changing capacitance, he suggested, some time ago (Cole and Curtis (6) p. 666) that conduction is confined to a small fraction of the total membrane area—the so called ion "channels." Further, his voltage clamp concept (Cole (5)) has been a spectacular tool that has been used in ingenious ways to obtain ion conductance information on nearly every kind of membrane preparation available. As Dr. Cole has noted (15), (postscript), the era of description of events from a point of view external to the membrane is drawing to a close and the present challenge is a description of events within the membrane.

A relatively new approach to how ions move through cell membranes is the measurement and analysis of fluctuation phenomena (noise) in small patches of membrane. For many years, the study of fluctuation phe-

[1] This work was supported in part by National Institutes of Health Grant NS 09857 and NS 11764.

nomena has been an important field in physics, particularly with respect to semiconductor materials and devices. In recent years, a small group of investigators has come to realize that noise concepts and measurement techniques, which have been developed in engineering and physics, could be applied to conduction problems in biological membranes. Noise measurements together with powerful concepts of fluctuation phenomena from statistical mechanics and information theory (14) can provide a basis for comparing models of ion conduction directly with experimentally derived statistics that relate to membrane microscopic processes.

Presently, there is a handful of investigators, of whom I am aware, who are engaged in membrane noise measurements. These include the Verveen group (Verveen and Derksen (22) Siebenga et al. (20)) in Leiden who are working on node of Ranvier in single frog fibers, Poussart (18) in Quebec who has studied lobster axons and who together with L. E. Moore has joined my own activities (9, 11) on squid axon. A recent addition to squid axon noise researchers is a group in Camogli. Although the techniques and results of the Camogli group (Wanke et al. (23), DeFelice (7)) differ from those that I shall describe, it is too early to resolve these differences. In addition to axon noise, Katz and Miledi (15) and Anderson and Stevens (1) have studied noise induced by acetylcholine in the frog neuromuscular junction.

POWER DENSITY SPECTRUM

The basic tool in characterizing noise is Fourier analysis (Blackman and Tukey (2). In Fig. 1 a two-terminal electrical network, N, is shown with terminal voltage V and current I. N is assumed to consist of a combination of dissipative and energy storage elements. If V and I are further assumed to be in a steady state, a mean value for V and I can be determined. In addition, with sufficient low noise amplification, fluctuations (ΔV, ΔI) in the time domain about the mean values of V and I can be measured as indicated in Fig. 1. In most physical systems the spontaneous fluctuations or noise can be represented or approximated as Gaussian, stationary, random processes with zero averages, by removing the mean value, so that all of the relevant statistical properties are contained in the frequency spectrum, which is obtained by a Fourier transformation of the time domain waveform into the frequency domain. The power density spectrum, $S(f)$, of either the voltage fluctuation waveform during constant mean current or the current fluctuation waveform during constant mean voltage may be obtained, where the mean square value of either ΔV or ΔI per unit measurement bandwidth is conveniently plotted versus frequency in a double log plot (Fig. 1). Although it is not the only means of characterizing noise, the power spectrum is the most understandable, and, therefore, the most useful. It immediately conveys how energy is distributed among the sinusoidal (Fourier) components that make up the fluctuation waveform, produced by the physical processes modeled by the elements in network N. In general, the power spectrum of voltage fluctuations, $S_V(f)$, is related (Fig. 1) to the power spectrum of current fluctuations, $S_I(f)$, through the square of the magnitude of the impedance, $|Z(f)|^2$, of network N. Furthermore, if we imagine N as representing only potassium conduction as contained in the Hodgkin-Huxley formulation, the current noise spectrum relates (Fig. 1) to fluctuations in potassium conductance, $G(f)$, as well as to fluctuations in the ion driving force, $(V-E_K)$, and maximum potassium conductance, $\bar{g}_K$. Fluctuations in the latter two terms should produce significant spectral features only at very high frequencies, and thus low frequency power spectra of current fluctuations should reflect directly fluctuations in ion channel conduction.

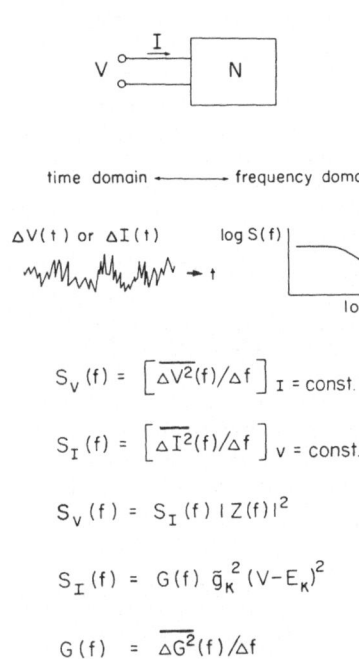

time domain ⟵⟶ frequency domain

$\Delta V(t)$ or $\Delta I(t)$ log S(f)

⟶ t

log f

$$S_V(f) = \left[\overline{\Delta V^2}(f)/\Delta f\right]_{I = const.}$$

$$S_I(f) = \left[\overline{\Delta I^2}(f)/\Delta f\right]_{V = const.}$$

$$S_V(f) = S_I(f)\,|Z(f)|^2$$

$$S_I(f) = G(f)\,\bar{g}_K^2\,(V-E_K)^2$$

$$G(f) = \overline{\Delta G^2}(f)/\Delta f$$

Figure 1. *Upper*: steady-state fluctuation waveform $\Delta V(t)$ or $\Delta I(t)$ at the terminals of network N (a combination of linear dissipative and energy storage elements) after amplification and removal of the mean. Power density spectrum, $S(f)$ versus f, conveys how energy is distributed among the sinusoidal components contained in the fluctuation waveform. *Lower*: the power spectrum during constant applied current ($I = $ const) is the mean square voltage fluctuation, $\overline{\Delta V^2}(f)$, at frequencies f per measurement bandwidth, Δf, and during constant applied voltage ($V = $ const) is the mean square current fluctuation, $\overline{\Delta I^2}(f)$, per measurement bandwidth. The voltage-noise power spectrum, $S_V(f)$, reflects both the current-noise power spectrum, $S_I(f)$, and the square of the magnitude of the impedance, $|Z(f)|^2$, of network N. In the HH axon with only K^+ conduction, $S_I(f)$ at low frequencies reflects predominantly "channel" fluctuations, $G(f)$, whereas at relatively high frequencies the spectrum reflects "open channel" conductance, $\bar{g}_K$, and driving force, $(V-E_K)$, fluctuations.

NOISE AND ITS RELATION TO ION CONDUCTION

A significant approach in establishing relationships between conduction and fluctuation phenomena in membranes arises from a consideration of a generalized shot-noise process (Papoulis (17)), see Fig. 2. From the point of view of a signal transmission scheme, membrane conduction may be thought of as a noiseless processor or filter (Fig. 2) with impulse response, h(t), which operates on an input function, a Poisson distribution of impulses, to produce an output function, a measurable spontaneous noise. The Poisson distributed impulses have as a physical basis the random motion of ions in the bulk solutions that cause ions to impinge on the membrane interfaces. The spontaneous noise resulting from ion motion, which relates to the previously mentioned fluctuation in driving force and maximum channel conductance, contains frequency components that extend to very high frequencies (MHz) and the information contained in these noises probably is not measurable by ordinary electrophysiological techniques. However, membrane conduction of ions may also modify or produce significant noise, i.e., the relaxation of relatively large molecular structures (symbolized by the noise source in Fig. 2) which could occur as part of or become the dominant feature of the measurable output noise. Noise of this type would have energy concentrated at relatively low frequencies and therefore should be measurable. An important point is that the spectral properties of low frequency noise are dependent on the impulse response of the membrane filter. At the level of spontaneous fluctuations, the amplitude domain is that of a "small signal" where linear analysis is assumed to apply. The complex Fourier transform, H(s), of the impulse response is simply the membrane complex impedance. Thus, the "small signal" membrane impedance becomes an important reference in the interpretation of noise data.

Three examples of power density spectra of output noise which result from various impedances are shown in Fig. 2. In an ohmic conduction process, the spectrum has the form $(\sin \theta/\theta)^2$, which reflects the spectrum of the input Poisson process. Note that this spectrum is flat, i.e., power is distributed uniformly over all frequencies, thus approximating "white" noise, up to the roll-off frequency. For an RC, the spectral form is Lorentzian, viz. $[1 + (2\pi f\tau)^2]^{-1}$. That is, the RC acts as a first-order filter with time constant τ for the input impulse sequence, which is assumed to approximate white noise to frequencies much greater than $(2\pi\tau)^{-1}$. Similarly, for an RLC circuit, which is underdamped, the spectrum shows peaking or an indication of resonance. The RC and RLC are reasonable first approximations of subthreshold impedance data on squid axon membrane (Cole (5) p. 84). With respect to measurements of membrane impedance, spontan-

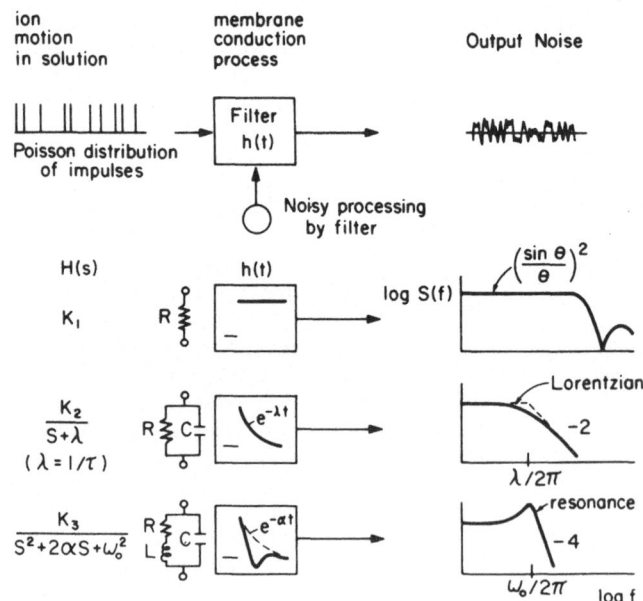

Figure 2. Membrane conduction noise conceptualized as a generalized shot process as described in text.

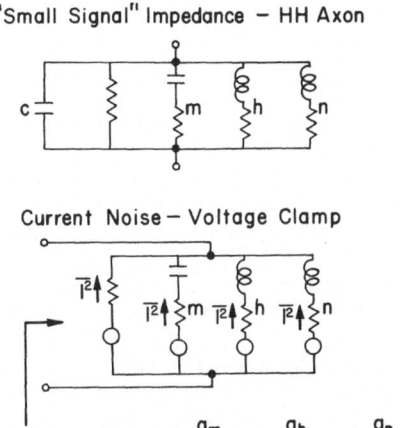

"Small Signal" Impedance – HH Axon

Current Noise – Voltage Clamp

$$Y(s) = g_\infty + \frac{g_m}{1+\tau_m s} + \frac{g_h}{1+\tau_h s} + \frac{g_n}{1+\tau_n s}$$

Figure 3. "Small-signal" equivalent circuit of axon membrane calculated by Cole ((5): Fig. 3.39) from linearized HH equations. Current noise contributions during voltage clamp can arise from every resistive element in the circuit (indicated by the $\overline{i}^2$ arrows). Complex ionic conductance, $Y(s)$, contains four terms, three of which are frequency dependent.

eous fluctuations can be complemented by the use of white noise as a stochastic input signal to determine membrane impedance and admittance. The cross power density spectrum, which contains both magnitude and phase information, is obtained through the Fourier transform of the cross-correlation of output with white noise input (Lee (16) p. 342). Such measurements of complex impedance and admittance have been made in squid axon membrane (11, 12).

A more detailed representation of the kinetics of ion conduction is contained in the complete "small-signal" impedance of axon membrane (Fig. 3) which was calculated from linearized HH equations[2] (Chandler, FitzHugh and Cole (3); Cole (5), p. 299). The frequency dependence of the familiar m, h and n parameters is

shown as resistances in series with reactive elements that are in parallel with a lumped resistor representing contributions from all ion conduction processes and, in addition, membrane capacitance. All resistive elements in this circuit are potential sources of noise that could contribute to membrane noise. Under voltage clamp conditions (Fig. 3) the membrane capacitive effects can be removed and current fluctuations reflect the remaining conduction processes. The complex admittance, $Y(s)$, of this circuit contains four terms, three of which are frequency dependent. If the elements of this circuit are noiseless except for the assumed uncorrelated "white" noise sources shown, the power spectrum of current fluctuations should have the same form as the sum of the squared magnitude of each of the admittance terms, since the current from each of the branches add to give the total current under constant voltage. The overall spectrum can be obtained by direct computation of the magnitude of each of the complex admittance terms from the linearized HH equations. Figure 4 shows the result of computations of the complex ionic-conductance (membrane capacitance excluded), $Y(s) = g' + jg''$, locus of the HH membrane at rest and several depolarized potentials from rest. I am indebted to Dr. Cole and Dr. Richard FitzHugh for supplying me with the parameters that enabled me to compute these curves with a desk calculator. At rest (0 mV), the admittance shows inductive reactance behavior at low frequencies due to the n process and then at high fre-

---

[2] The Hodgkin-Huxley equations are referred to as HH equations. The terms HH axon, HH membrane and HH power spectra all refer to use of the HH equations as an analog to compute, simulate or predict conditions in real axons.

quencies shows capacitance reactance behavior due to the m process. At 10 and 20 mV depolarization, the real part of the admittance shows negative conductance behavior as the m process becomes dominant, as Cole found before the voltage clamp in his calculations (Cole (5) Fig. 1.51), of the unpublished 1941 data taken with Marmont, and at 30 mV the real part of the admittance becomes positive again as n and h take over. Notice the large change in the magnitude of the admittance, i.e., the vector from the origin to any same frequency point on each of the curves.

Figure 5 shows the sum of squared magnitude terms of ionic admittance, $|Y(f)|^2$, as a function of frequency at rest potential and 10, 20, 30 mV depolarizations from rest. These spectra (heavy lines) result from the superposition of the squared magnitude of each of the four terms in the admittance expression in Fig. 3. The frequency dependences of the four individual components are also shown, from which the relative contributions can be assessed. There are three important spectral features in Fig. 5. At low frequencies the spectrum is dominated at all potentials by the n relaxation process. An additional higher frequency relaxation from the m process is evident at rest potential, most obvious at 10 mV, and disappears at 30 mV. Finally at high frequencies the spectrum approaches a level determined by g_∞^2. These spectra represent a qualitative indication of the frequency dependence of the current noise power spectra based on the linearized HH equations if the noise of all dissipative elements in the circuit of Fig. 3 is modeled by white noise sources. The form of the resultant noise spectrum depends on the actual magnitude of the sources. With the assump-

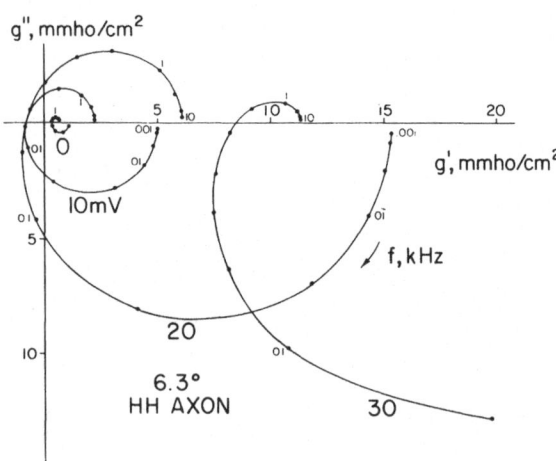

Figure 4. Cole-Cole plot of the complex ionic conductance, $Y(s) = g'(f) + jg''(f)$, calculated from linearized HH equations at rest potential (0 mV) (Cole, (5): Fig. 3.43) and for 10, 20, 30 mV depolarizations from rest. Note low frequency inductive reactance ($g'' < 0$) due to n and high frequency capacitive reactance ($g'' > 0$) due to m process, negative values of g' at 10 and 20 mV where g_m becomes significant, and large increase in $|Y|$ with depolarization.

HH AXON 6.3°

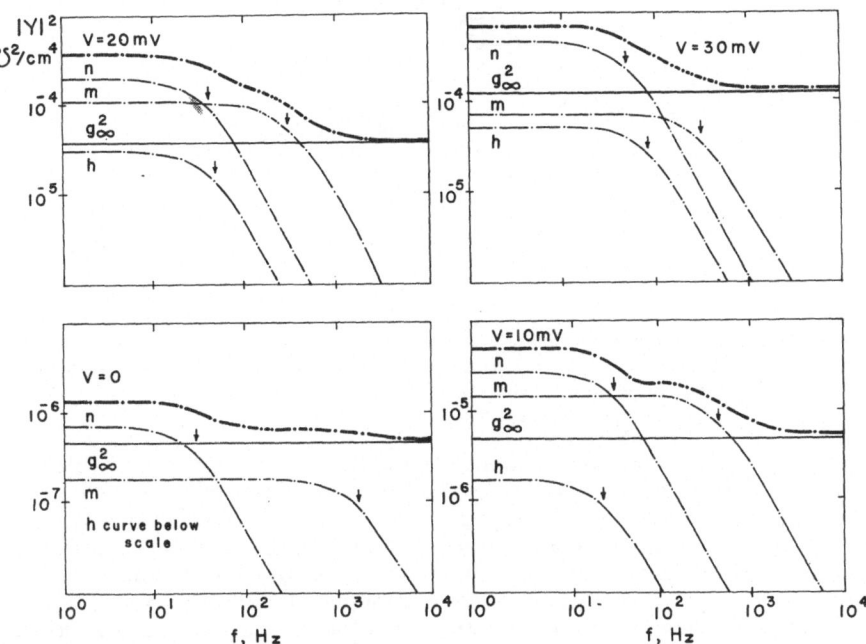

Figure 5. Sum of the squared magnitude of the complex ionic conductance terms versus frequency (heavy curves) computed from the linearized HH admittance (Fig. 3) which results from the superposition of the squared magnitude of each of four processes m, h, n and g_∞ shown separately (light curves). Note that the n relaxation process dominates the spectrum at low frequencies, the m relaxation is most obvious at $V = 10$ mV, and $g_\infty{}^2$ determines the high frequency level. Arrows indicate corner frequency of Lorentzian spectra.

tion of white noise sources, the admittance spectra may be used for comparison and as a guide to conductance fluctuation spectra in real axons.

PRACTICAL
NOISE MEASUREMENTS

There are two steps in implementing membrane noise measurements, the acquisition of the membrane noise "signal" with a minimum of extraneous noise, and the subsequent analysis of noise to obtain rapid and accurate power spectra. Each step can be accomplished in many different ways. I shall confine my comments

to a few important principles, and then discuss briefly the methods we have used (9, 11). The foremost consideration in measuring membrane noise is the area of membrane from which fluctuations are to be measured. A practical and essential matter concerns the minimization of extraneous amplifier noise in the measurement of membrane noise (Poussart (18, 19)). The noise properties of any amplifier can be characterized by an equivalent voltage noise and current noise source at its input (Cherry and Hooper (4)) as shown in Fig. 6. In general, these sources are frequency dependent.

Low Noise Amplification

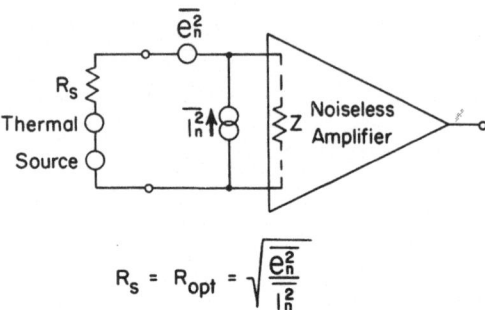

$$R_s = R_{opt} = \sqrt{\frac{\overline{e_n^2}}{\overline{i_n^2}}}$$

Figure 6. Characterization of the noise properties of an amplifier in terms of intrinsic voltage and current noise sources that are dependent on frequency. In most instances, these sources can be assumed to be uncorrelated.

The best signal/noise performance is obtained from an amplifier when the source resistance of the signal being measured properly matches the optimum noise resistance, R_{opt}, defined as the square root of the ratio of intrinsic voltage-to-current noise power of the amplifier. It should be emphasized that R_{opt} is not the resistance for maximum power transfer and is frequency dependent. Field-effect-transistor (FET) input amplifiers provide the best signal/noise performance of any device in the frequency range of 1–1,000 Hz with source resistances in the range of 1–10 MΩ. Thus, if optimum signal/noise is desired from an FET amplifier, the patch resistance must be from 1 to 10 MΩ. For a 1,000 $\Omega \cdot$cm^2 squid axon, the signal/noise condition requires[3] measurement from a patch area of 10^{-3} to 10^{-4} cm^2. The requirement of a small membrane area also applies to measurements of current fluctuations during voltage clamp since the clamp must be capable of responding to patch voltage-fluctuations in order to correct for them and maintain a constant voltage condition.

In the analysis of noise to produce power spectra there are many methods. The two which offer the most spectral detail as well as the best statistical accuracy are 1) implementation of a Fourier transform program by digital computer, and 2) signal time compression. The latter method is the one we use because of its speed and compactness (9, 11). If a signal is sampled periodically and the digitized samples are stored sequentially in a circulating memory, the original signal can be reconstructed from the digital sequence in memory in a much shorter time interval than originally required. This process is called time compression. It preserves the spectral content of the original signal, but translates all of the frequency components to very high frequencies, where analog filter-

---

[3] It should be noted that Wanke et al. (23) report noise measurements from very large areas (0.053–0.36 cm^2) of axon membrane (3–40 kΩ) with an FET amplifier. From the point of view of proper source resistance for low noise operation from an FET amplifier as well as minimization of contributions from electrode noise, such low source resistances are far from optimal. For example, they measure a power density of 4.4×10^{-23} A^2/Hz from a 3.1 kΩ resistor at 300 Hz which is an order of magnitude above the thermal level (5.2×10^{-24} A^2/Hz at 20 C).

ing and detection of the components can be done quickly and with high resolution. This technique is particularly useful in producing audio and subaudio frequency "real-time" spectra where usual analog filtering techniques require very long times (minutes) to obtain estimates of a single frequency component. In contrast, time compression instruments produce a several hundred point spectrum in about 0.1 sec. Thus many "real-time" spectra obtained from the same time series can be averaged digitally to give significant improvements in statistical accuracy, especially when dealing with fluctuating signals.

PATCH ISOLATION

The method which we use for electrical isolation of small patches of squid axon is shown in Fig. 7. Two glass pipettes are drawn and assembled in a co-axial configuration. The inner pipette has a 60–100 μm diameter tip and is filled with sea water. It also has a floating Pt-Pt wire within its length to reduce the pipette equivalent AC resistance to 10 kΩ (10). The tip of the inner pipette is brought into contact with the outer axon surface and isosmotic sucrose solution is introduced into the outer pipette, which directs the flow of sucrose over the portion of axon that surrounds the axon underneath the aperture of the inner pipette. In this way isolation resistances of 1 MΩ or more are usually obtained. Patch action potentials can be elicited. From measurements of patch capacitances of 10–100 pF (depending on inner pipette tip diameter) and assuming a membrane capacitance of 1 μF/cm^2, patch area is estimated to be 10^{-4} to 10^{-5} cm^2. The patch can be voltage clamped by inserting a potential electrode in the path leading to the patch and making the appropriate connections to a clamp system. Figure 8 illus-

trates patch spikes and voltage clamp current records. There obviously are numerous tests to assure adequacy of this voltage clamp technique as well as low noise operation of the patch technique. These tests have been made and are (11) or will be described elsewhere.

NOISE RESULTS

Figure 9 shows power density spectra of voltage fluctuations from a single patch of squid axon membrane. These data were obtained from spontaneous fluctuations in an excitable patch at the rest potential and for prolonged hyperpolarizations and prolonged small depolarizations from rest. The power levels are two

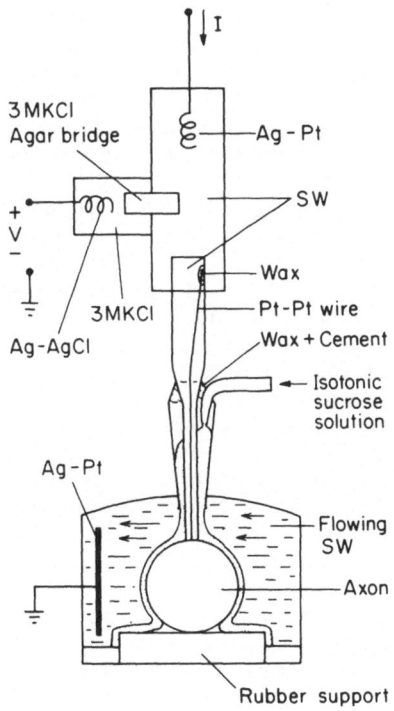

Figure 7. Scheme for electrically isolating a patch of squid axon membrane with sucrose solution.

orders of magnitude in excess of thermal noise. The spectral component that produces the pronounced humps[4] is suggestive of noise associated with a relaxation process (9). The corner frequencies, indicated by arrows, that were obtained from the intersection of extrapolated low and high frequency asymptotes are clearly voltage and temperature dependent in a way reminiscent of the n process relaxation calculated from the HH power spectra and shown in Fig. 5. However, before a comparison with the predicted HH power spectra can be made, several points must be established: 1) since current noise spectra reflect conductance fluctuations directly whereas voltage noise spectra are, generally, modified by membrane impedance (see relations in Fig. 1), the equivalence of current and voltage spectra must be demonstrated; and 2) the relation of noise spectra to specific ion movements must be demonstrated by changes in spectra when alterations in ion conduction are made.

With respect to the first point, Fig. 10 compares voltage noise spectra with current noise spectra during voltage clamp from the same axon patch, at rest and for polarizations of patch potential from rest (11). The two sets of spectra, which have been superposed, show relatively good agreement. The major difference is in the rate of decline at high frequencies, which is greater in voltage spectra. Notice, however, that the corners and rest of the spectra are nearly the same. These data suggest that patch membrane impedance does not alter significantly patch voltage fluctuations, in which case (see relations in Fig. 1) patch voltage

fluctuations would reflect current fluctuations that in turn reflect conductance fluctuations. An explanation of this condition arises from consideration of the patch parameters. Patch area has been estimated from capacitance measurements to be 10^{-4} to 10^{-5} cm^2. For a 1,000 Ω cm^2 membrane this gives a patch resistance of 10–100 MΩ. However, the apparent patch resistance, i.e., patch resistance in parallel with sucrose solution isolation resistance, is measured typically to be 1 MΩ. Thus the 1 MΩ is almost entirely a measure of the shunt resistance since the patch resistance is greater by an order of magnitude or more.

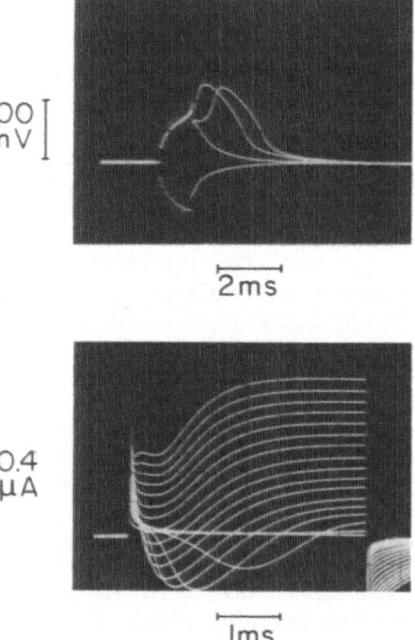

100 mV

2ms

0.4 µA

1ms

Figure 8. *Upper*: patch spikes elicited by constant current pulses through patch. *Lower*: patch voltage-clamp currents during step changes of potential. The ohmic shunt current through the sucrose solution which "isolates" the patch has been removed electronically. 100 µm (ID) pipette. 10 C.

---

[4] Similar humps have been observed in other preparations by Siebenga et al. (20) and Anderson and Stevens (1).

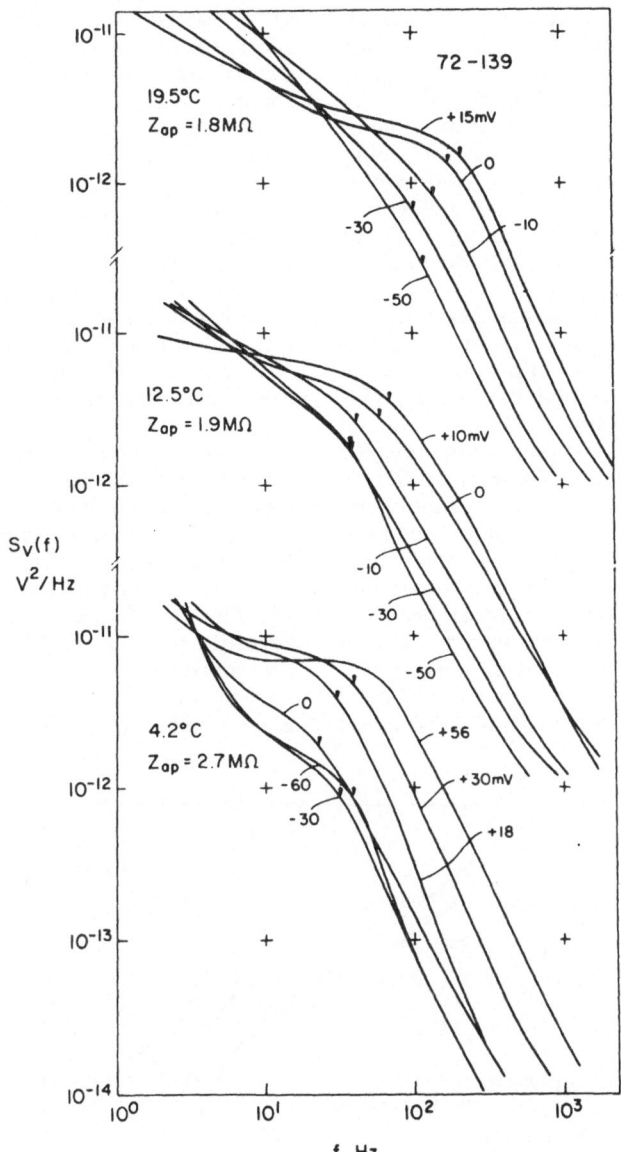

Figure 9. Voltage-noise power spectra from a single patch of squid axon for polarizations from rest potential (0 mV) and at three different temperatures.

Consequently, the well-known impedance resonance that occurs in squid axon (5) is well damped by the relatively low shunt in the patch isolation technique which produces an impedance function that is flat to 100 Hz or more and then declines monotonically (elimination of the impedance resonance in squid axon has been demonstrated with a membrane resistance to shunt resistance ratio of 5:1, Fishman, Poussart and Moore (11)). The damped patch impedance then produces only high frequency filtering of voltage noise and insofar as spectral features occur below several hundred Hz, patch voltage spectra reflect current fluctuations (9).

This hypothesis has been tested, as shown in Fig. 11, by considering Cole's first approximation of membrane impedance—the RLC circuit. In addition, a "white" voltage noise source is placed in the RL branch, and a resistor is added in parallel with the RLC to model the effect of a finite patch isolation. With values chosen that relate to 1 cm² of membrane area and a patch resistance to isolation resistance ratio of 5:1, the spectra in part (a) of Fig. 11 are obtained as C is varied. The $C = 0$ condition is equivalent to a voltage clamp in which the effect

of C is removed. The resulting spectrum is a pure Lorentzian, produced by the conversion of the "white" voltage noise by the RL impedance into an equivalent Lorentzian-distributed current noise source, which drives current into the 200 Ω shunt. The current noise produces voltage fluctuations across the 200 Ω resistor that have the same spectral form as the RL impedance, which models conduction kinetics. The effect of adding C is merely to add and increase the filtering of the Lorentzian spectrum produced by the RL branch. The same filtering effect was noted in the comparison between current and voltage noise spectra from an axon patch in Fig. 10. The effect of various values of isolation resistance on voltage noise spectra is shown in part (b) of Fig. 11. The actual spectra are displaced vertically from one another; however, they have been superposed for comparison of form. In this case, 1 μF is assumed as the membrane capaci-

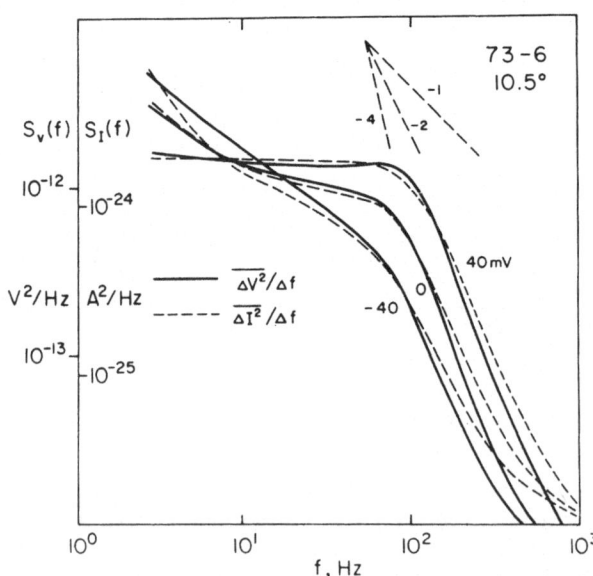

Figure 10. Comparison of voltage and current-noise power spectra from the same patch.

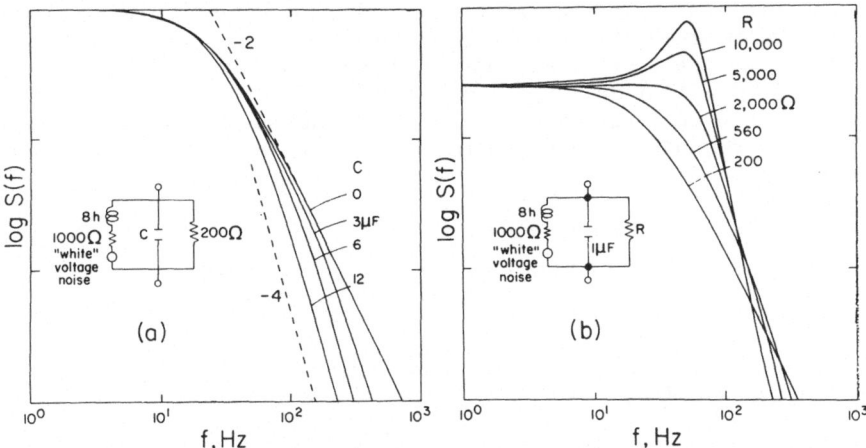

Figure 11. Power spectra of voltage noise measured across the terminals of the circuits in (a) and (b) which model the patch measurement with an RL branch, representing the kinetics of a single conduction process, a C element representing patch capacitance, and an R element, representing the finite isolation resistance through the sucrose solution, and a "white" noise source to produce fluctuations in the RL branch. The circuit elements were chosen to correspond to 1 cm² of squid axon. In (a) the spectra are for various values of C while all other elements were constant and the shunt R value was chosen to correspond to a patch-resistance to shunt-resistance ratio of 5:1. The spectrum for $C = 0$ is a Lorentzian and corresponds to the spectral form that is measured regardless of the value of C if the circuit in (a) is voltage clamped and current noise is analyzed (since V-clamp removes the effect of the C element). Addition and increase of capacitance merely produces high frequency filtering, i.e., an increased rate of decline, which reaches a slope of -4 for $C = 12$ μF, while the corner frequency and low frequency portion of the voltage spectrum are not significantly altered. In (b) the capacitance was held constant at 1 μF and the shunt R varied. The spectra for this case have been superposed to compare form. For $R = 200$ Ω, a good Lorentzian is obtained up to frequencies beyond 100 Hz. As the shunt resistance approaches and exceeds the value of the impedance of the RL branch, peaking occurs and the voltage-noise power spectrum no longer reflects directly the current noise of the RL branch.

tance. With $R = 200$ Ω, a good Lorentzian is obtained well past the corner. Only when the RL impedance becomes comparable to or less than the impedance of the parallel combination of patch capacitance and isolation resistance does the spectrum depart from Lorentzian behavior to give enhanced sharpness at the corner and a steeper high frequency decline. Eventually, peaking, which is a manifestation of an underdamped RLC circuit, becomes evident. These measurements together with the current and voltage noise spectra from the same patch thus suggest the near

equivalence of patch voltage and current spectra, provided that the impedance of the parallel combination of isolation resistance and patch capacitance is relatively low compared to the patch conduction impedance.

With respect to the relation of noise to specific ion processes, Fig. 12 shows results on voltage noise spectra produced by blocking K^+ currents with tetraethylammonium (TEA) ion (8). The control spectra with "humps" were obtained from a patch of axon internally perfused with buffered potassium fluoride solution. Since active transport is eliminated by KF-perfusion, the controls indicate an insensitivity of these fluctuations to active transport. However, after addition of 50 mM TEA the noise component which produces the spectral humps is eliminated, leaving only residual noise spectra that, when subtracted from the controls at each potential, give apparent Lorentzian spectra.

Tetrodotoxin in concentrations of 10^{-6} M was ineffective in producing

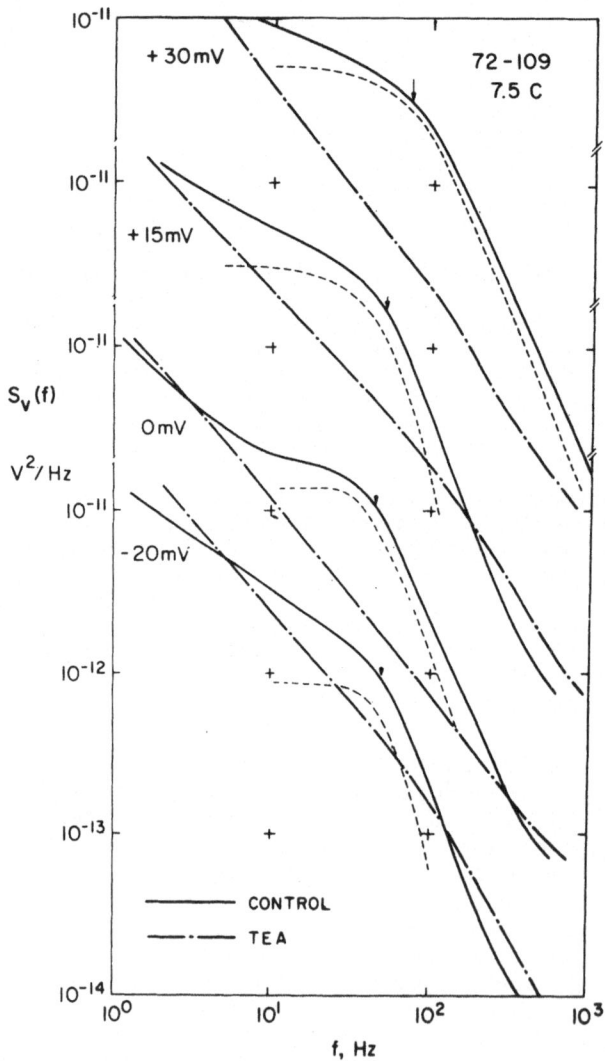

Figure 12. Voltage noise spectra from a patch of axon which was internally perfused with buffered KF solution (controls) and after perfusion with 50 mM TEA added (interrupted curves). Dashed curves were obtained by graphically subtracting TEA curves from controls. (From Fishman (9).)

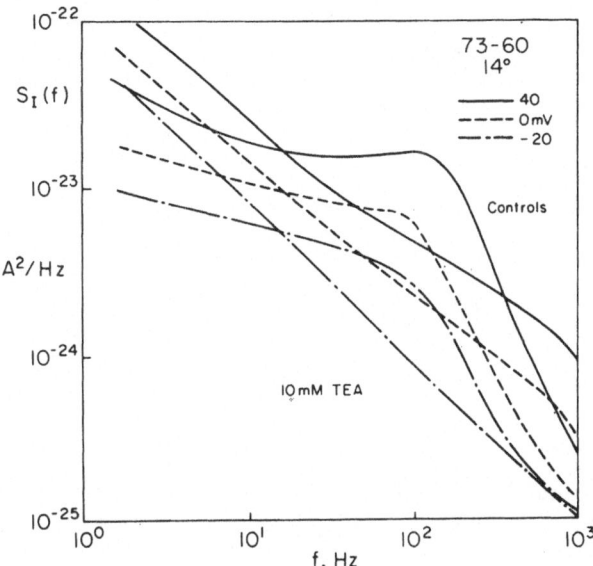

Figure 13. Current-noise power spectra during patch voltage clamp for the same conditions as described in Fig. 12 except with 10 mM TEA.

changes in the apparent Lorentzian spectra which suggests that this noise component does not relate to Na$^+$ conduction. Figure 13 shows data from essentially the same kind of experiment except that current noise from voltage-clamped patches has been analyzed (11). The current noise spectra corroborate the dependence of the Lorentzian noise component on K$^+$ current flow since the humps again disappear after addition of TEA. These data as well as the previous spectra which show dependence on V and T suggest strongly that the Lorentzian noise component relates to voltage-dependent potassium-ion conduction.

With respect to Na conduction noise, positive identification of Na$^+$-related noise components has not been made in any of our noise spectra from patches in well over 300 axons, although spectral components other than the Lorentzian ones do occur (11). At present, we have not observed

a relaxation in noise spectra that would correspond to Na$^+$ noise predicted by the HH conduction admittance. It would appear that special agents that keep Na channels open for relatively long times are required.

Finally, a few words about the expectations of noise measurements. They offer, perhaps, the best available counterpart to measurements of the first order statistics of individual conducting channels in lipid films. The second order statistics, i.e., noise from populations of conducting channels in axon patches, may not be as elegant, but potentially they are as definitive. The use of noise data as a spectroscopic tool (8) to obtain microscopic information on ion-channel kinetics can provide a direct basis for comparing and relating models of conduction. For example, Fig. 14 shows data of spectral corner frequencies versus potential and temperature for K$^+$ noise measurements (9) compared

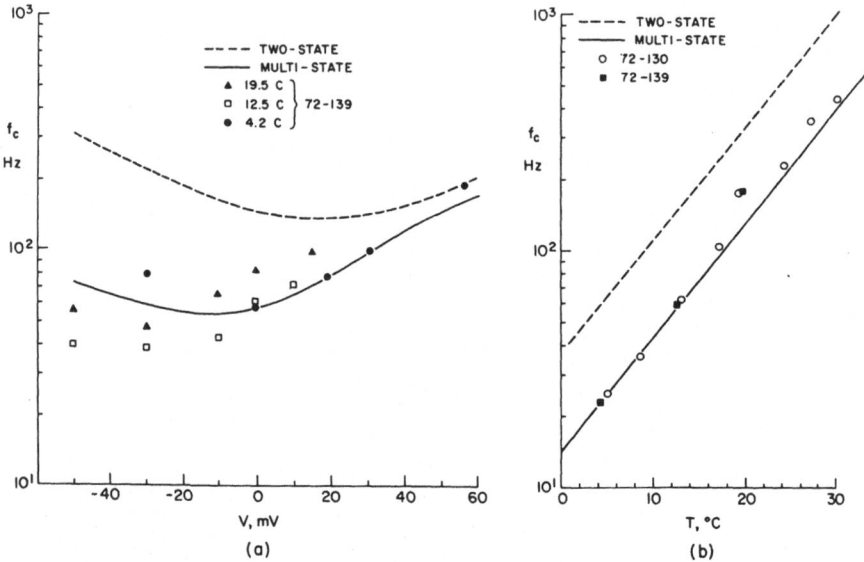

Figure 14. Comparison of corner frequency versus potential (*a*) and temperature (*b*) from patch noise spectra with prediction from HH two-state K$^+$ conductance power spectra computed by Hill and Chen (13) and Stevens (21). (From Fishman (9).)

with theoretical curves generated from HH power spectra of two-state K$^+$ channel conductances, derived independently by Hill and Chen (13) and Stevens (21). The validity of many contemporary interpretations, based on macroscopic conduction data, rests on resolution of the crucial but largely ignored issue of two-state versus multistate conduction. Refinements in the analysis of conduction fluctuations in addition to analysis of mean conduction characteristics may lead to resolution of this and equally important questions.

REFERENCES

1. ANDERSON, C. R., AND C. F. STEVENS. *J. Physiol. London* 235: 655, 1973.
2. BLACKMAN, R. B., AND J. W. TUKEY. *The Measurement of Power Spectra.* New York: Dover, 1958.
3. CHANDLER, W. K., R. FITZHUGH AND K. S. COLE. *Biophys. J.* 2: 105, 1962.
4. CHERRY, E. M., AND D. E. HOOPER. *Amplifying Devices and Low-Pass Amplifier Design.* New York: Wiley, 1968.
5. COLE, K. S. *Membranes, Ions and Impulses.* Berkeley, CA: Univ. of California Press, 1968.
6. COLE, K. S., AND H. J. CURTIS. *J. Gen. Physiol.* 22: 649, 1939.
7. DEFELICE, L. J. *Federation Proc.* 34: 1338, 1975.
8. FEHER, G., AND M. WEISSMAN. *Proc. Natl. Acad. Sci. U.S.* 70: 870, 1973.
9. FISHMAN, H. M. *Proc. Natl. Acad. Sci. U.S.* 70: 876, 1973.
10. FISHMAN, H. M. *IEEE Trans.* BME-20: 380, 1973.
11. FISHMAN, H. M., D. J. M. POUSSART AND L. E. MOORE. In preparation.
12. GUTTMAN, R., L. FELDMAN AND H. LECAR. *Biophys. J.* 14: 941, 1974.
13. HILL, T. L., AND Y. CHEN. *Biophys. J.* 12: 948, 1972.
14. JAYNES, E. T. *Phys. Rev.* 106: 620, 1957.
15. KATZ, B., AND R. MILEDI. *J. Physiol London* 224: 665, 1972.
16. LEE, Y. W. *Statistical Theory of Communication.* New York: Wiley, 1960.
17. PAPOULIS, A. *Probability, Random Variables*

and Stochastic Processes. New York: Mc-Graw-Hill, 1965.

18. POUSSART, D. J. M. *Biophys. J.* 11: 211, 1971.

19. POUSSART, D. J. M. *Rev. Sci. Inst.* 44: 1049, 1973.

20. SIEBENGA, E., A. W. A. MEYER AND A. A.

VERVEEN. *Arch. Ges. Physiol.* 341: 87, 1973.

21. STEVENS, C. F. *Biophys. J.* 12: 1028, 1972.

22. VERVEEN, A. A., AND H. E. DERKSEN. *Proc. IEEE* 56: 906, 1968.

23. WANKE, E., L. J. DeFELICE AND F. CONTI. *Arch. Ges. Physiol.* 347: 63, 1974.

Potassium and sodium current noise from squid axon membranes[1]

LOUIS J. DeFELICE[2], E. WANKE AND F. CONTI

*Department of Anatomy, Emory University, Atlanta, Georgia 30322
and Laboratorio di Cibernetica e Biofisica, Camogli (Genova), Italy*

ABSTRACT

Current noise under voltage clamp has been measured from the giant axon of *Loligo vulgaris*. Relatively large areas (up to 0.4 cm²) were used for these measurements, under standard space clamped conditions. Current noise spectral densities were studied in the range of 0 to 20 C, −100 to −40 mV and in normal, tetrodotoxin, and tetraethylammonium ion solutions. Three components of current noise were found, l/f-noise, and a K and a Na component.—DeFELICE, L. J., E. WANKE AND F. CONTI. Potassium and sodium current noise from squid axon membranes. *Federation Proc.* 34: 1338–1342, 1975.

The experimental work to be described was done during the spring and summer of 1973 at the Laboratorio di Cibernetica e Biofisica in Camogli, Italy. The data were subsequently analyzed at Emory University. One publication (10) has resulted from this work. Figure 1 is taken from that paper. Other preliminary results have been presented (3) and a complete manuscript is in press (2). The work presented here is selected material from this larger study.

THEORETICAL CONSIDERATIONS

We have used a preparation with relatively large membrane area. We first address the question "can noise be measured from large membrane areas?" The answer divides naturally into two parts, one concerned with the

---

[1] Supported by the National Science Foundation and the Consiglio Nazionale delle Ricerche.

[2] Supported by the Georgia Heart Association and the Research Corporation during parts of this study.

47

source of noise and the other with the device used for measuring noise.

The source of noise is in this case an isolated patch of excitable squid axon membrane. (Details of the isolation, an air gap technique, are already published (10)). The source presents itself as a two terminal device for which one may define a complex small-signal impedance. Noise from the membrane may be modeled either as a voltage source, a current source, or both. We do not know beforehand the correct model to use for an excitable membrane.

Consider the membrane to be a parallel arrangement of voltage sources (e) each with internal resistance (r). The mean square voltage noise expected from such a system is

$$e_s^2 = \frac{\alpha^0}{\alpha} e^2 \ \text{V}^2/\text{Hz} \qquad (1)$$

where α^0 is the effective area of a single generator (or $1/\alpha^0$ is the surface density of the generators e) and α is the total area of the membrane. Consider instead a parallel arrangement of current sources (e/r for each branch). The mean square current noise expected from such a system is

$$i_s^2 = \frac{\alpha}{\alpha^0} \frac{e^2}{r^2} \equiv \frac{\alpha}{\alpha^0} i^2 \ \text{A}^2/\text{Hz} \qquad (2)$$

The assumption in both *equation 1* and *2* is no correlation of the individual sources. Statistically dependent sources would give a factor $(\alpha^0/\alpha)^0$, or its inverse.

α^0, e and r are constant state properties of the membrane (here taken as homogeneous). Voltage spectral density is inversely proportional, while current spectral density is proportional to membrane area, as regards the source of noise.

Consider the measurement of this noise. Amplifiers are active, three terminal devices whose contribution

to the total noise depends not only on its internal sources but on the source impedance presented to the input of the amplifier.

Consider first the measurement of voltage. A popular model (6,7,9), here somewhat simplified, predicts the measured open circuit voltage from a source (or a parallel array of sources e) to be

$$V = e_s + e_n + i_n Z \qquad (3)$$

where $e_n(\text{V}/\sqrt{\text{Hz}})$ is the amplifier voltage noise, $i_n(\text{A}/\sqrt{\text{Hz}})$ is the amplifier current noise and $Z\ (\Omega)$ is the source impedance.

The signal-to-noise ratio for voltage spectral density is (Z^2 implies $|\bar{Z}|^2$ throughout)

$$\left(\frac{S}{N}\right)_v = \frac{(\alpha^0/\alpha)e^2}{e_n^2 + i_n^2 Z^2} \qquad (4)$$

Let

$$Z = Z_m/\alpha \qquad (5)$$

where $Z_m(\Omega \cdot \text{area})$ is a property of the membrane of area α, and is a (complex) function of frequency. Then

$$\left(\frac{S}{N}\right)_v = \frac{\alpha^0 \alpha e^2}{\alpha^2 e_n^2 + i_n^2 Z_m^2} \qquad (6)$$

This function of area has a maximum when α has the value

$$\alpha_{max} = \frac{i_n}{e_n} Z_m \qquad (7)$$

Consider now the measurement of current. The same amplifier model, again with certain simplifications, predicts

$$-I = \frac{V}{R_f} = i_s + i_n + \frac{e_f}{R_f} + \frac{e_n}{R_f} + \frac{e_n}{Z} \qquad (8)$$

where R_f is the feedback resistor and e_f is its associated voltage noise. Suppose that the two terms involving R_f in *equation 8* are negligible. Then the signal-to-noise ratio for current spectral density is

$$\left(\frac{S}{N}\right)_I = \frac{(\alpha/\alpha^0)i^2}{i_n{}^2 + \alpha^2(e_n{}^2/Z_m{}^2)} \quad (9)$$

This function of area also has a maximum value when α is given by *equation 7*.

Thus, for either a voltage or a current measurement (regardless of how the source is modeled) there is a best area to use for noise measurements. Furthermore, this area is the same. This result depends on the assumptions implied in *equations 1* and *2*, and on the model used for the amplifier. (For example, if the elementary sources were perfectly correlated with one another, *equation 7* would be an inequality defining not a best area but an area, beyond which the voltage signal-to-noise ratio would continue to improve.)

Reasonable values for a field-effect-transistor (FET) input amplifier at 60 Hz are

$$e_n{}^2 \cong 10^{-16} \text{ V}^2/\text{Hz} \quad (10)$$

$$i_n{}^2 \cong 10^{-28} \text{ A}^2/\text{Hz} \quad (11)$$

Membrane properties, of course, are highly variable. Normal squid axon membranes have strong temperature, transmembrane voltage, bathing solution, and frequency dependence. Choose $T = 8$ C, transmembrane potential of -60 mV, artificial sea water bathing solution at 60 Hz (1). Then

$$Z_m{}^2 \cong 10^7 \Omega^2 \cdot \text{cm}^4 \quad (12)$$

The above values give

$$\alpha_{max} = 3 \times 10^{-3} \text{ cm}^2 \quad (13)$$

The value of the voltage signal to noise ratio at α_{max} is

$$\left(\frac{S}{N}\right)_v^{max} = \frac{\alpha^0 e^2}{2i_n e_n Z_m} \quad (14)$$

The value of the current signal to noise ratio at α_{max} is

$$\left(\frac{S}{N}\right)_I^{max} = \frac{(1/\alpha^0)i^2 Z_m}{2i_n e_n} \quad (15)$$

These two expressions are identical as may be seen from the definitions of the voltage and current sources given by *equations 1* and *2*, and from *equation 17*.

The value of α_{max} given by *equation 13* is not to be taken literally. It depends on both the source and on the amplifier and it may be varied. Because of the frequency dependence of Z_m, there is not one value of α_{max} for the entire spectrum. Also, α_{max} is voltage and temperature dependent, making it clearly impossible to work at this optimal area in any realistic experiment. The dependence on the amplifier involves the ratio i_n/e_n over which we have some control. For either by selection of the input transistor, varying its operating point, or by a suitable combination of input transistors, the ratio i_n/e_n may be varied at will. For example, a parallel combination of equal transistors halves $e_n{}^2$ and doubles $i_n{}^2$ to increase α_{max} by a factor of two. Altering i_n and e_n in this way does not effect the value of the signal to noise ratio, since there, only the product $e_n i_n$ is involved.

What is the penalty for being at an area other than α_{max}? Choose

$$Q = \frac{\Delta\alpha}{\alpha_{max}} \quad (16)$$

as a measure of the broadness of the ratio (S/N), where $\Delta\alpha$ is the breadth of (S/N) at half-maximum. It can be shown that on the low side of α_{max}, $Q = 0.7$, and on the high side, $Q = 2.7$. In other words, working at an area $\frac{1}{3} \alpha_{max}$ has about the same degrading effect on (S/N) as working at 4 α_{max}. The ratio (S/N) against α is asymmetric about α_{max}, varying more slowly for larger areas.

We have selected our input amplifiers to be compatible with large areas 0.04 to 0.4 cm². This was done in order to achieve a simple preparation, and to insure against damage, poor clamping or otherwise adversely affected axons. Amplifier noise is present in our measurements. Its effect is shown in Figs. 3 and 4. Two and sometimes four FET inputs in parallel were used. Selection was based on obtaining sufficiently low values of $e_n^2/|Z|^2$ at high frequencies.

We have measured voltage noise, current noise and small signal impedance from large membrane areas, as nearly simultaneously as possible. These are related by (8, 10)

$$e_s^2 = i_s^2 Z^2 \text{ V}^2/\text{Hz} \qquad (17)$$

One expects the voltage and the current spectral densities to be different since Z is a strong function of frequency in the normal squid axon membrane. For example, near rest (about -60 mV) Z^2 shows a resonant frequency which varies between 30 and 100 Hz as the temperature varies between 3 C and 14 C (1). Z^2 is also strongly voltage dependent, varying from a simple RC-like behavior at hyperpolarizations to a much more complicated resonant behavior at depolarizations (5). The simple capacitive component of Z, dominant at frequencies above about 300 Hz, would be expected to make the voltage and current spectral densities different.

Consider a source of noise represented by several branches in an equivalent circuit of the membrane. For example, e_1, Z_1, and e_2, Z_2 in parallel. Then

$$e_s = \frac{e_1 Z_2 + e_2 Z_1}{Z_1 + Z_2} \qquad (18)$$

and

$$i_s = \frac{e_1 Z_2 + e_2 Z_1}{Z_1 Z_2} \qquad (19)$$

Calculating the densities, assuming e_1 and e_2 independent, we note that the numerators would be equivalent but

that the denominators would have, in general, a rather different frequency dependence. (e_1^2, and e_2^2 may have any spectral shape at all.) In this way the above arguments can be extended to heterogeneous systems.

METHODS

A more detailed account of the experimental set up is given in ref. 10. Briefly, giant axons from *Loligo vulgaris* were used in an external air gap isolation technique (the internal air gap technique also discussed in ref. 10 does not represent any data in the present paper), very similar to the standard space-clamp preparation. Lengths of axon 2 cm long and 450–600 microns in diameter were isolated. The artificial sea water had the composition, in units of mM/liter, of 10 KCl, 450 NaCl, 40 $MgCl_2$ and 10 $CaCl_2$. Tris-HCl buffer was used to adjust the pH to 7.9. The amplifier consisted of a FET input stage, the details of which will be given in a subsequent publication (2). A 61 KΩ feedback resistor was used in the voltage clamping circuit. Three to four minutes of data were taken from each membrane state. Figure 1 uses a narrow band analogue filter method to estimate the spectra. The rest of the current-noise data were analyzed digitally on an H-P 5451 A Fourier Analyzer. A hanning window was used on data uniformly sampled at 2,000 times per second and assembled into blocks of 1,024 points. This implies a record length of 0.512 sec and a frequency resolution of 1.95 Hz. Each spectrum represents an average of 100–300 individual spectra of 512 points, 1.95 Hz apart.

RESULTS

Figure 1 is from an earlier paper (10). It verifies *equation 17* for squid giant axon membrane near its resting state. Z^2 was measured by injecting sinu-

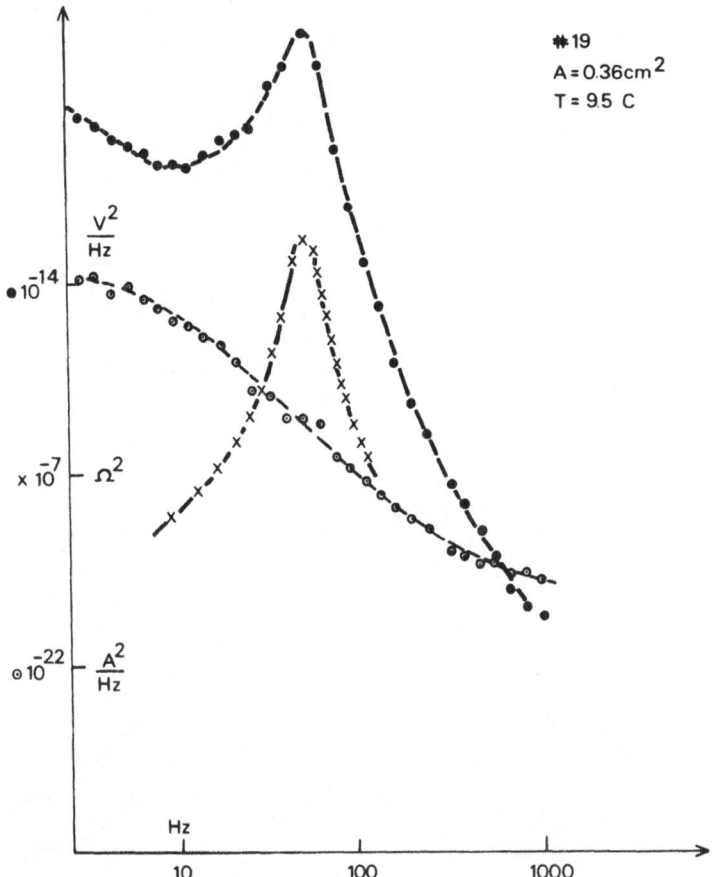

Figure 1. Current noise spectral density (⊙), voltage noise spectral density (●), and the square of the modulus of the small signal impedance (×) from squid axon membrane at rest. The dashed curves are simply smooth lines drawn through experimental points. (From ref 10.)

soidal currents, and measuring the voltage response across the membrane, kept below 1 mV. This property of Z^2 follows directly from the standard equations describing this membrane (5). The remaining figures show current spectral densities under voltage clamp exclusively.

Figure 2 shows the normal dependence of current noise spectral densities on mean membrane voltage (shown as a parameter in mV). Near rest (−57 mV) a minimum occurs, obvious at lower frequencies. On hyperpolarization (−96 mV) a 1/f-noise component becomes evident. On depolarization (−47 mV) a knee becomes evident in the densities. 50 Hz and 150 Hz peaks from the line voltage are visible on most of the curves.

Figure 3 shows a current spectral density of a membrane near rest (−56 mV) in greater detail. A smooth line is drawn through the spectral curve. Also shown are three small-signal impedance curves, (Z^2), one near rest (-56 mV), one at a hyperpolarized level (-82 mV) and the third at a depolarized level (-49 mV). The -56 mV curve gives a value for Z at the low frequency limit, for which the Johnson current noise density $4kT/R$ is shown ($4kTReZ/Z^2$ might have been used). The curve just above this is the dominant amplifier noise term e_n^2/Z^2, plus the Johnson noise term, shown for our measured value of e_n^2 and the Z^2 at -56 mV. Ampli-

fier noise becomes important at the high frequencies due mainly to the behavior of Z^2, and at low frequencies due mainly to the behavior of e_n^2. At depolarized levels one expects the low frequency spectra to contain a greater error due to amplifier noise. The solid dots show the subtraction of the amplifier noise from the measured current spectral density.

Figure 4 shows the current spectral density for a slightly depolarized axon treated with tetrodotoxin (TTX) (2×10^{-8} M/liter added to the external solution). The solid line

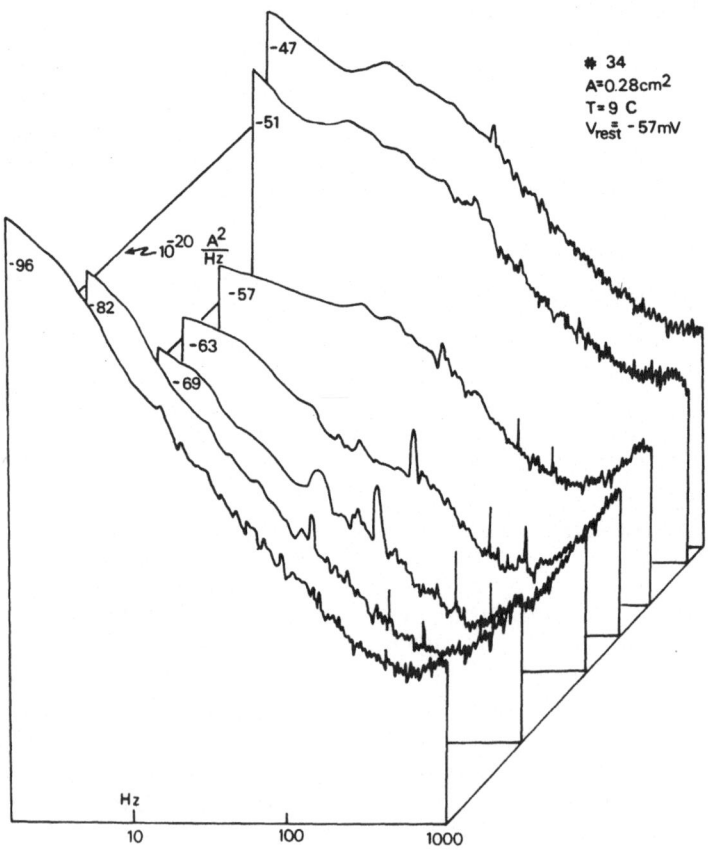

Figure 2. Current noise spectral densities from one axon at fixed temperature with membrane potential in mV shown as a parameter.

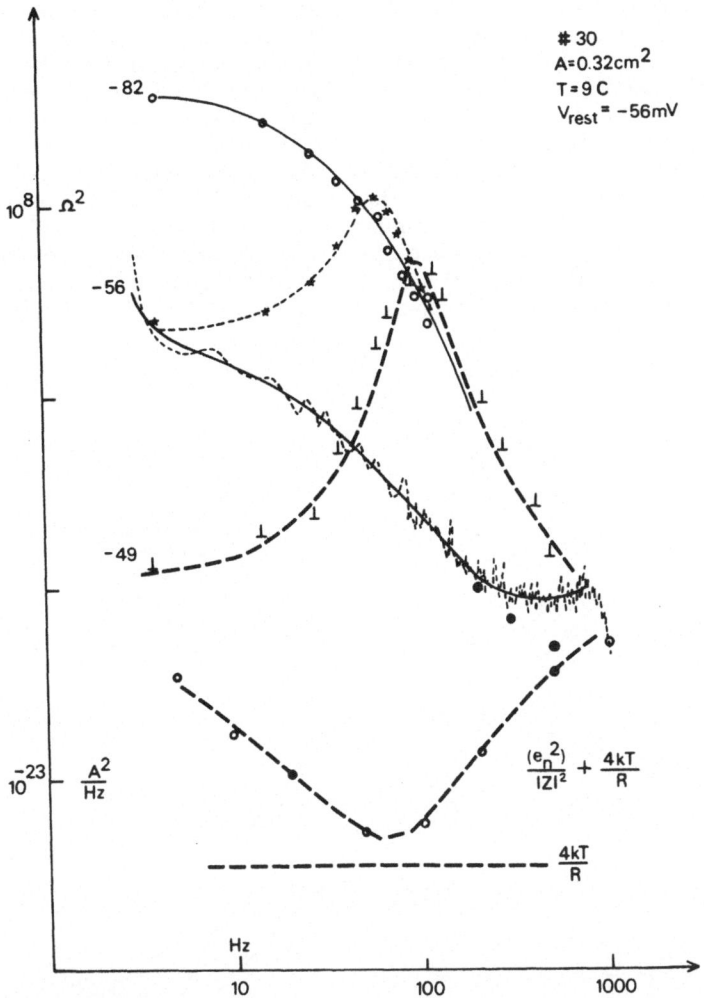

Figure 3. Current noise spectral density of a membrane at rest. A continuous line is simply drawn through the experimental points. The bottom dashed line is the spectrum 4kT/R, where R is taken from the low frequency of the small signal impedance measurement at rest ($\star$). Impedance curves are also shown for hyperpolarized ($\bigcirc$) and depolarized ($\perp$) levels. The curve immediately above 4kT/R is the amplifier noise expected from the measured amplifier voltage noise and the measured resting impedance.

through the spectrum is the sum of a pure b/f component and a $c/(1 + \omega^2\tau^2)$ component (called a Lorentzian), both of which are also shown. The values of b (at 1 Hz) and c, both in A^2/Hz, are given in the figure. $f_c = 38$ Hz refers to the cut-off value $f_c = 1/2\pi\tau$, and therefore $\tau = 4.2$ msec. The subtraction of amplifier noise is shown at high frequencies. Note the failure of the sum of the l/f and the Lorentzian component to fit the data in this region. The fact takes on significance below, when a third spectral component is discussed.

Figure 5 shows results from two axons. One shows the spectrum in artificial sea water (ASW) and then in TTX-ASW. The second shows the spectrum for tetraethylammonium ion (TEA) (70 mM/liter applied internally). All the data are on the same absolute scale and have been selected

to have the area, temperature and membrane potential nearly the same (the hyperpolarizing effect of TTX and the depolarizing effect of TEA are reflected in the values of the clamping currents and mean voltages, also shown in the figure). The curves labeled TTX and TEA nearly sum to the curve labeled ASW. Note especially the behavior above about 100 Hz. This figure is central in the argument that both K and a Na current

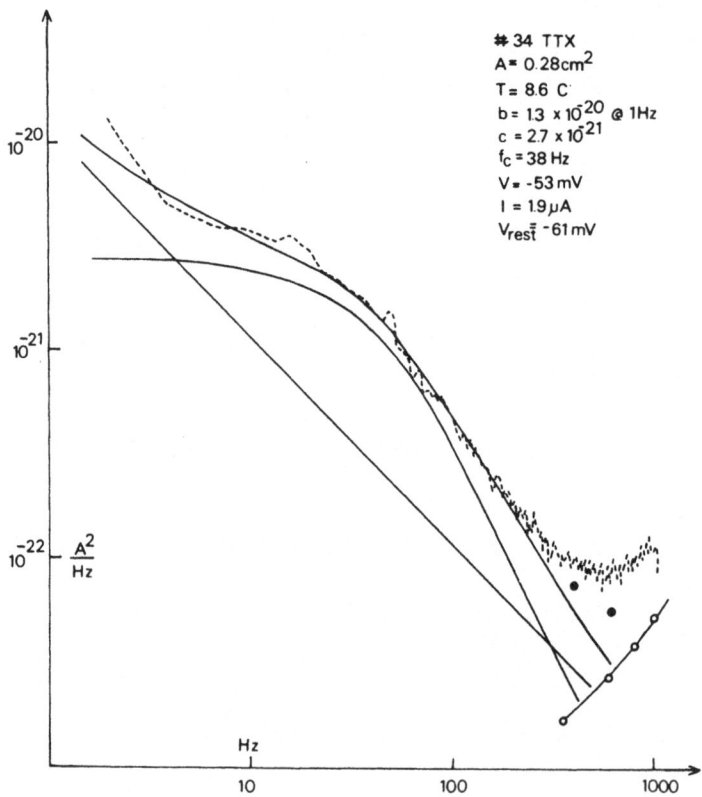

Figure 4. Current noise spectral density from a depolarized TTX-axon. The solid line through the experimental points is the sum of a b/f and $c/(1 + \omega^2\tau^2)$ component, both shown beneath it. The parameters b, c and f_c are given in the graph $f_c = 1/2\pi\tau$. The open points are measured amplifier noise from a passive analogue circuit mimicking the high frequency membrane impedance at this potential. The filled points are the difference of the axon and the passive circuit noise measurements.

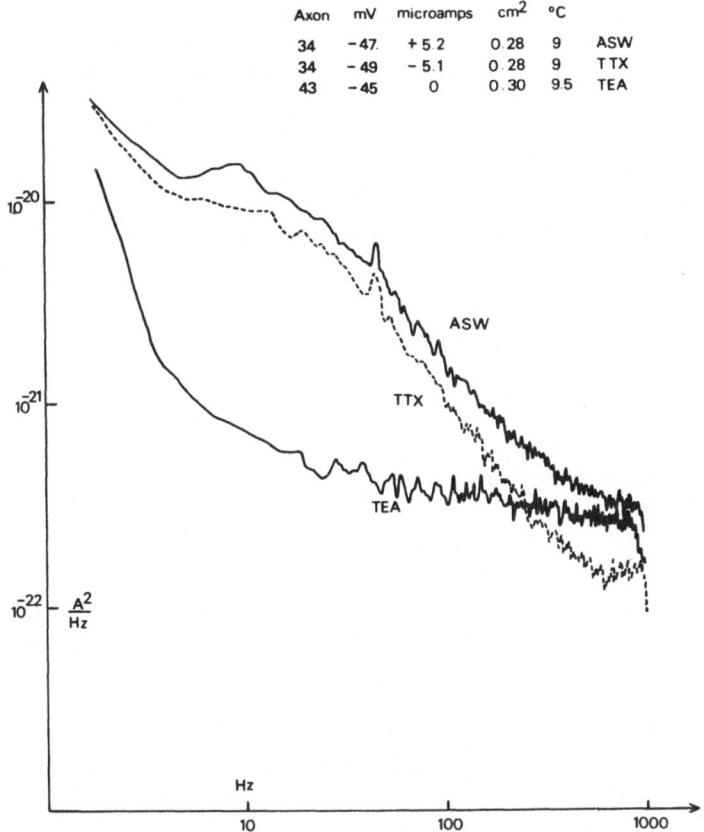

Axon	mV	microamps	cm²	°C	
34	−47	+5.2	0.28	9	ASW
34	−49	−5.1	0.28	9	TTX
43	−45	0	0.30	9.5	TEA

Figure 5. Three current noise spectral densities from two comparable axons. The same absolute scale applies to all three cases. The effect of TTX and TEA on the normal spectrum (ASW) for this voltage and temperature state is demonstrated.

noise component have been detected in the normal squid axon membrane.

Figure 6 shows the value of c (the low frequency limit) for the Lorentzian spectrum versus membrane potential. The procedure shown in Fig. 4 was used to extract values of c. The upper group of points are from 8 ASW and TTX axons between 8 and 10 C, normalized to 1 cm². The lower group of points are c-values from a second Lorentzian that may be drawn in TTX and TEA experiments at higher frequencies. These data and the procedures are given in detail in ref. 2. They represent here a rather broad temperature range of 4 to 18 C.

DISCUSSION AND SUMMARY

The emphasis in this report is on the practical notion that noise may be measured from larger areas than have previously been attempted, and on the separation of the measured current noise into spectral components.

The assumption of uncorrelated noise sources in the membrane leads naturally to a best area at which to work. The reason for this is that the membrane contributes as α while the amplifier contributes as α^2.

The signal-to-noise ratio is maximal at some area, and this area is the same for either a voltage or a current measurement. The value of this best area depends on the ratio i_n/e_n, and may therefore be adjusted since amplifiers may easily be ar-

ranged to increase or decrease one of these at the expense of the other. The value of the ratio S/N, the same for either a voltage or a current measurement, is not affected by this procedure since it involves the product $e_n i_n$. Of course, the actual value of S/N will depend on how noisy the source is; in the squid axon

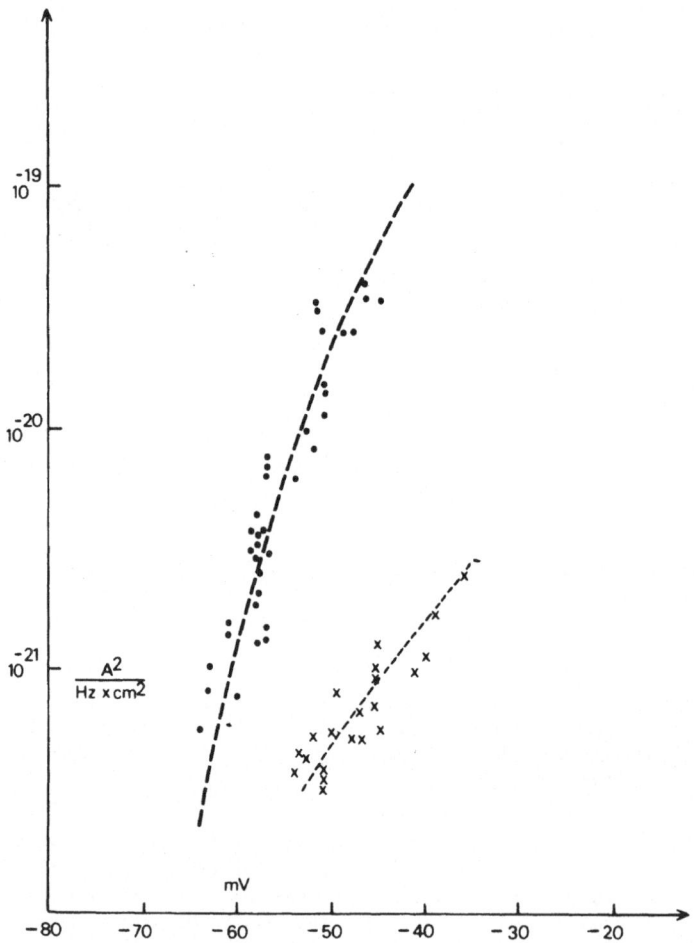

Figure 6. The voltage dependence of the low-frequency limit of two Lorentzian current noise components extracted graphically after the method of Fig. 4. The data for K ($\bullet$) are taken from eight axons in the range 8 to 10 C and normalized to 1 cm² of axon mem-

brane, and include ASW and TTX experiments. The data for Na ($\times$) are taken from three TEA and three TTX axons at temperatures in the range of 4 to 18 C and normalized to 1 cm² of axon membrane.

membrane it is sufficiently noisy to make these experiments possible with the membrane area and the amplifier parameters we have selected.

In any case, since the optimal area also involves $Z_m(\omega)$, there is no one best area for the entire frequency range of the spectral density. Also, the optimal area depends on voltage and temperature.

S/N is an asymmetric function about the optimal area, varying more slowly with area at larger values than at smaller values. We have chosen to work at relatively large areas, and the measured amplifier noise (see Fig. 3) supports this choice. This noncorrelation method limits our accuracy as we approach 1,000 Hz, due primarily to e_n^2 and the capacitive component of the membrane's impedance. The upward swing in some current spectral densities approaching 1,000 Hz may therefore be regarded as entirely spurious.

Current spectral densities under voltage clamp are less affected by normal resonance and the capacitive component of Z. This does not imply that they are completely unfiltered by the membrane's impedance. We have shown that the current spectral density consists of at least three components. One is the well-known 1/f noise, present at all membrane potentials but more obvious at hyperpolarizations. It must be carefully subtracted to observe the second component, a Lorentzian, more obvious at depolarizations.

A third component, less accurately measured than the other two, was also modeled as a Lorentzian. Its presence is suspected only after TTX and TEA spectra are compared with normal spectra. Only with low temperatures could the beginning of the roll-off be seen in our frequency range.

Tetrodotoxin has its effect primarily at the high end of our spectra. It is interpreted as removing current noise due to the membrane's Na system. The difference in slopes of the ASW and TTX spectra of Fig. 5 above 100 Hz reflects this loss. This interpretation is supported by the TEA spectrum. Tetraethylammonium ion has its effect primarily at the lower end of the normal spectrum. It is interpreted as removing current noise due to the membrane's K system. Figure 5 suggests a second Lorentzian due to Na, whose roll-off is not seen at this temperature and in this frequency range. This interpretation is supported by the fact that the low frequency limits of the two noise components are both voltage dependent, as shown in Fig. 6. A full discussion is given in ref 2.

Our experiments are not accurate enough to decide between true Lorentzian and the near-Lorentzian shapes predicted by others (4, 8). Nevertheless, the low frequency limits of the K and the Na noise components reported here should be extremely useful in testing these models and in measuring fundamental membrane properties, such as the unitary conductance of K and Na channels and their surface densities. This aspect of the work is pursued in a later publication (2) along with the dependence of τ and 1/f noise on temperature and voltage.

CONCLUSIONS

a) Voltage and current noise may be measured accurately from relatively large areas of squid axon membrane, considerably larger than an optimal area suggested by a simple theory.

b) Current noise is less affected by the membrane's impedance than is voltage noise, and is more eaily related to theories of membrane fluctuations.

c) At least three components of squid axon membrane current noise exist. The first is 1/f noise; the second is a relaxation spectrum related to the

K-system, and the third is a relaxation spectrum related to the Na-system. The K-system has a larger characteristic time associated with it than the Na-system, at a given temperature. The K-system has a larger magnitude at lower frequencies than the Na-system, at a given voltage.

REFERENCES

1. CONTI, F. Nerve membrane electrical characteristics near the resting state. *Biophysik* 6: 257–270, 1970.
2. CONTI, F., L. J. DEFELICE AND E. WANKE. K and Na current noise in space clamped squid giant axon. *J. Physiol. London* In press.
3. DEFELICE, L. J., E. WANKE AND F. CONTI. K and Na current noise in nerve membrane. *Federation Proc.* 33: 1339, 1974.
4. HILL, T. C., AND Y. CHEN. On the theory of ion transport across the nerve membrane IV. Noise from the open-close kinetics of K channels. *Biophys. J.* 12: 948–959, 1972.
5. MAURO, A., F. CONTI, F. DODGE AND R. SCHOR. Subthreshold behavior and phenomenological impedance of the squid giant axon. *J. Gen. Physiol.* 55: 497–523, 1970.
6. POUSSART, D. Membrane current noise in lobster axon under voltage clamp. *Biophys. J.* 11: 211–234, 1971.
7. POUSSART, D. Low level average power measurements: noise figure improvements through parallel or series connection of noisy amplifiers. *Rev. Sci. Instr.* 44: 1049–1052, 1974.
8. STEVENS, C. F., Inferences about membrane properties from electrical noise measurements. *Biophys. J.* 12: 1028–1047, 1972.
9. VERVEEN, A. A., AND L. J. DEFELICE. Membrane noise. *Progr. Biophys. Mol. Biol.* 28: 189–265, 1974.
10. WANKE, E., L. J. DEFELICE AND F. CONTI. Voltage noise, current noise and impedance in space clamped squid giant axon. *Pflügers Arch.* 347: 63–74, 1974.

Voltage clamp simulation[1]

J. MAILEN KOOTSEY

Department of Physiology and Pharmacology
Duke University Medical Center, Durham, North Carolina 27710

ABSTRACT

The voltage clamp experiment on the squid giant axon is successful because of the special characteristics of the preparation: cylindrical shape, large diameter, and so on. The method is much more difficult to apply to small cells and to networks of cells because voltage gradients and unwanted stray impedances are not readily eliminated. Simulation of the voltage clamp experiment is proposed as a method for determining when these factors and the characteristics of the clamp electronics affect the experimental results, for evaluating experimental techniques for improving the quality of the clamp, and as a possible method of learning something about the membrane when no experimental improvement is feasible. The numerical methods for including one spatial variable in the analysis are reviewed briefly. Several examples of voltage clamp simulations are discussed: double sucrose gap clamp of axons, clamp of the giant synapse, single sucrose gap clamp of cardiac muscle bundles, point clamp of the end of a fiber bundle, and the steady-state three-microelectrode clamp of a cable with nonlinear membrane. The results indicate that the quality of a clamp cannot be evaluated from the voltage and current records as commonly measured.—KOOTSEY, J. M. Voltage clamp simulation. *Federation Proc.* 34: 1343–1349, 1975.

In 1955, Kacy Cole (4) reviewed the early development of the squid axon voltage clamp and described it as having "required a novel and unique biological approach coupled with keen insight and prodigious effort." The outstanding success of the experiments in accounting for the electrical properties of nerve membrane was due in part to a careful match of the preparation (squid axon) and the theoretical model (one-dimensional cable equation). The method still would not have been useful, however, without the special experimental maneuvers designed to eliminate troublesome derivatives in the cable equation. As Kacy said later in his review, when we "meet a limit to our powers of analysis of an involved situation, the only thing to do is to simplify the situation."

Since 1955, the highly successful voltage clamp technique has been

[1] This work was supported by National Institutes of Health, NHLI Grants HL 12157 and HL 11307.

applied to a variety of other cells and tissues. In most, if not all, of these applications the preparation and theoretical model were not nearly as well matched as were the squid axon and the cable model. Furthermore, unwanted derivatives have been difficult or even impossible to eliminate by experimental maneuvers. Some experimenters have made serious efforts to minimize these problems while others have apparently ignored them. Looking over these experiments, it seems that we are now frequently finding ourselves in the reverse situation of the early days of axon voltage clamp. To paraphrase Kacy's words, when we "meet a limit to our (experimental) powers of simplifying the situation, the only thing to do is to expand our powers of analysis." Fortunately, such expansion has become reasonable because of the tremendous increase in availability of computational power. In particular, it is now possible to retain in the analysis some of the troublesome derivatives (e.g., the second spatial derivative in the one-dimensional cable model) as well as the characteristics of the apparatus such as amplifier gain, bandwidth, stray capacitances, leakage resistances, and the like. This paper will first present an overview of the technique for bringing one spatial variable back into the analysis, then will follow with some examples of the application of simulation to the analysis of voltage clamp experiments.

CABLE SIMULATION METHOD

The voltage clamp experiment on a one-dimensional cable is described by a set of differential equations, some of which may be nonlinear: a partial differential equation describing the variation of transmembrane potential in space and time, one or more ordinary differential equations de-

scribing the membrane properties, and one or more ordinary differential equations describing the clamp amplifier and other experimental apparatus. Given a set of parameter values and a desired wave form for the command voltage, the goal is to find solutions to this set of differential equations resulting in values for the transmembrane potential $V(x,t)$ along the length of the cable as a function of position x and time t as well as other voltages, currents, and conductances.

There is no general method for finding analytic solutions for such a system of nonlinear equations with arbitrary boundary conditions. Numerical methods (see, for example, 3 and 5) can, however, always be used to find solutions for V at discrete values of the position along the cable and time as well as to integrate the associated nonlinear ordinary differential equations. Figure 1A shows a typical integration grid of points for the cable in the x-t plane. Points along a horizontal row represent values of the transmembrane potential V at points along the cable at a fixed time (a "snapshot"). A vertical column represents transmembrane potentials at a fixed position as a function of time (such as might be recorded by a fixed electrode). Typical values of the length increment Δx range from a few μm to a few hundred μm and the value of the time increment Δt may be a few μsec or tens of μsec.

The rectangles along the bottom row represent points where the transmembrane potential V has a known initial value. The procedure is to calculate V at the points on the second row (at time Δt) from the known potentials according to the partial differential equation after it has been converted to discrete form. This conversion can be done in a variety of ways, relating V at different numbers

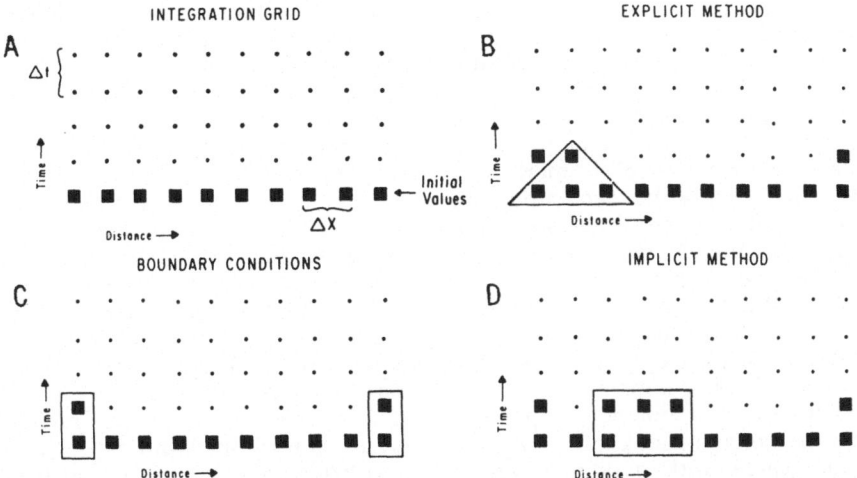

Figure 1. Diagrammatic illustration of numerical integration methods for solving the cable equation. A) grid of points in the x,t plane where the solution for V is desired. B) the triangle represents one application of the difference equation from an explicit method. C) the rectangles surround points where the potentials are related by boundary conditions. D) The rectangle represents one application of the difference equation from an explicit method.

of points and approximating the partial differential equation to different degrees of accuracy. The triangle in Figure 1B represents a simple explicit approximation, the partial differential equation reducing in this case to an algebraic relation between an unknown potential at one point to known potentials at three points. Repeated applications of this explicit equation give all but two of the unknown potentials on the second row. The remaining two potentials are calculated from equations that describe the conditions at each end of the cable (the "boundary conditions," Figure 1C), e.g., terminations, current injection, and so on. When all the potentials in the second row have been calculated, this row is treated as another set of initial conditions and a new row is calculated. As the solution for V is advanced in time, the associated ordinary differential equations for membrane variables, the

clamp amplifier, and the like are also integrated with the same time steps.

Accuracy in the solution of the voltage clamp equations generally demands a small time increment Δt and thus a large number of time steps to simulate a single clamp run. A careful choice of the method of converting the partial differential equation to discrete form can increase the accuracy for a given Δt and thus reduce the required number of time steps. An improved integration method is illustrated by the rectangle in Figure 1D. This is an implicit method and the resulting difference equation relates three known potentials to three unknown potentials. The difference equation is applied $N - 2$ times (where N is the number of grid points along a horizontal row) and when combined with the two boundary conditions, the result is a set of N linear algebraic equations in N unknowns that must be solved at each time step. Fortunately, the special nature of these equations results in an easily solved system where the amount of computation increases linearly with N rather than with N squared. The improvement in accuracy over the explicit method is such that the possible decrease in the number of steps required more than offsets the increased amount of computation per step.

Even with the best available numerical methods, the total amount of computation required to solve a cable equation is enormous. Consider, for example, the solution of the equations for the voltage clamp of a 100 segment cable with Hodgkin-Huxley membrane properties (6). Computing the solution for 100 msec would require about 10,000 time steps. All of the alphas and betas describing the membrane characteristics (each includes one or more exponentials) would have to be evaluated 1,000,000 times. One hundred equations in 100 unknowns would have to be solved 10,000 times and 300 or more ordinary differential equations would have to be integrated through 10,000 time steps each. This computation can be done in as little as a minute on a large computer.

CLAMP SIMULATION APPLICATIONS

Early in the 1950's, strange "notches" and oscillations were observed in the current records from squid axon voltage clamps. Some (15) interpreted these records as evidence of a mechanism different from that envisaged by Hodgkin and Huxley (6), but in 1960 Taylor, Moore and Cole (16) by a combination of measurements and simulation were able to show that these peculiarities in the current record were probably due to inadequate space clamp, i.e., inadequate elimination of the second spatial derivative. Because of the limited computational power available at that time, the spatial distribution of membrane had to be simulated by just two patches of membrane.

Recently, Moore, Ramon and Joyner (12) have simulated the double sucrose gap and other clamp configurations for squid and lobster axons. In their simulations, the axon was represented by a one-dimensional cable (21 segments) with membrane having active properties as described by Hodgkin and Huxley. Also included in the model were characteristics of the clamp amplifier (gain, bandwidth, input capacitance) and values of the external feedback components. Solutions to the equations were found by the methods outlined in the previous section. As expected, the simulation showed that it was possible to obtain a fast clamp of a one-dimensional cable to a uniform transmembrane potential when the node was very short. When the node length was increased to one third of the resting length constant, the maximum deviation in potential between the controlled end and the end where the current was injected was 12 mV. Figure 2 illustrates the results when the node length was increased to 1.2 length constants. Note that the difference in potential between the ends of the cable is as large as 40 mV for a 50 mV clamp step. The local current density at the ends of the cable undergoes oscillations similar to the experimental notches, but these are only barely visible on the record of total input current. The effect of resistance r_s in series with the membrane was also investigated. It was found that if r_s was 7 ohm-cm or more, its presence caused changes in the time course of currents and changes in the slope of the negative resistance region, but did not change the measured sodium equilibrium potential. Values of r_s

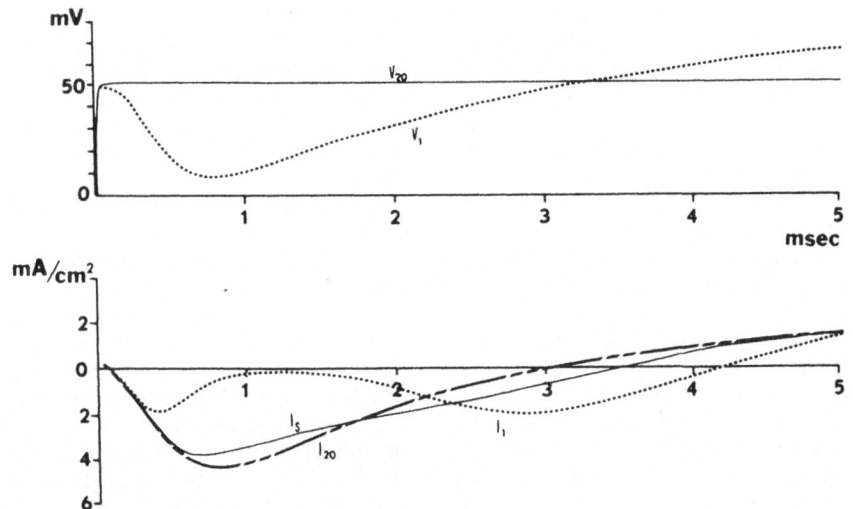

Figure 2. Results of simulation of double sucrose-gap voltage clamp of lobster axon of diameter 125 μm, length 500 μm (from Moore, Ramon and Joyner (12). V_1, I_1: membrane voltage, current at current injection end. V_{20}, I_{20}: membrane voltage, current at end where voltage is measured. I_S: total membrane current.

greater than 50 ohm-cm resulted in propagated action potentials after a clamp step, although the total current record showed no notches or other unusual behavior.

Joyner, Moore, and Ramon (9) have also simulated the voltage clamp of a short isolated length of the post-synaptic region of the squid giant synapse using a finite cable as a model of the preparation. They found that the sodium equilibrium potential could be determined quite accurately, but that the measurement of the reversal potential E_s for the synaptic conductance could be in error by 20 mV or more depending on the length of the isolated region, the positions of the microelectrodes, and the value of the membrane resistance. Blocking the active potassium conductance (e.g., with tetraethylammonium) was found to reduce the error in determining E_s.

There is little doubt as to the validity of the classical voltage clamp

method when properly applied to the squid axon. The contribution of simulation to squid axon clamps is in deciding on the interpretation to be given to small deviations between experiment and theory: Does the deviation call for a revision of theory or does it result from small deviations from the ideal experiment, such as nonzero series resistance or finite amplifier gain and bandwidth?

For other less ideal preparations such as bundles of cardiac cells, however, simulation is more of a necessity for the results of voltage clamp are always potentially contaminated with artifacts originating from imperfections of the method. Simulation would seem to be the only way of deciding which results must be attributed, until proved otherwise, to spatial and temporal inhomogeneities of membrane potential and which results can reasonably be attributed to the properties of the membrane. In other words, when spatial derivatives, leakage conductances, and so on, cannot be eliminated experimentally, the only recourse is to try to assess their effects by simulation. Conclusions from such a simulation can range from no effect on the results to complete ambiguity of the experiment. To illustrate the importance of simulation in studying nonideal preparations, and how it can interact with the experiments, I would like to follow an attempt to understand the mechanism of the cardiac action potential through several logical steps. In the following narrative, "we" refers to E. A. Johnson, M. Lieberman and the author.

In 1971, Johnson and Lieberman (7) reviewed the voltage clamp literature for cardiac preparations and criticized the papers published to date for their failure to recognize the probable deviations of the experiments from the classical "ideal clamp." They suggested, on the basis of qualitative and semiquantitative arguments, that the presence of factors such as spatial inhomogeneity of membrane potential and leakage meant that there were alternative explanations for the published results that were equally plausible as those given. To see if it was likely that anything could be learned about the early currents using presently available

techniques, we decided to try a detailed simulation (10) of the most popular clamp method—the single sucrose gap—including the effects of distributed membrane and the characteristics of the clamp electronics.

Figure 3A is a diagram representing our first model which included a clamp amplifier with finite gain and bandwidth and a one-dimensional cable to represent a column of cylindrical cells 10 μm in diameter. Since we did not have available a well-documented membrane description, we used a simple modification of the Hodgkin-Huxley model, a modification that produced a cardiac-like plateau and was consistent with unambiguous experiments on the early currents. We found this model to be very frustrating; there seemed to be no combination of amplifier gain and bandwidth that would result in clamp stability along with response anywhere near the speed of published experiments.

Our conclusion was that the model must be incomplete and that we must include other known factors about the preparation or experiment. We thus added to the model the three additional resistances shown in Figure 3B: the internal and external resistances

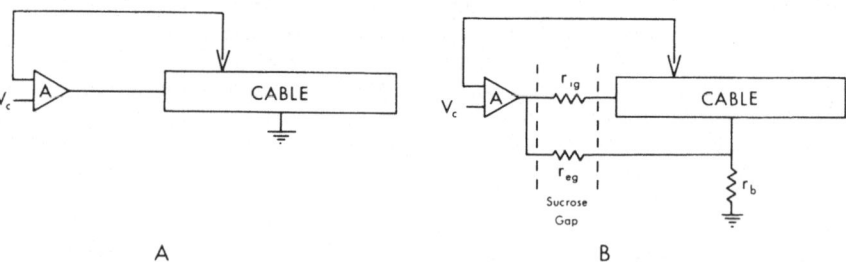

Figure 3. Models of the single sucrose-gap voltage clamp system (reproduced from 10). "CABLE" represents the preparation (assumed to behave electrically as a one-dimensional cable) and A is the clamp amplifier. The resistances r_{ig} and r_{eg} represent intracellular and extracellular resistances in the sucrose gap and r_b represents the membrane series resistance.

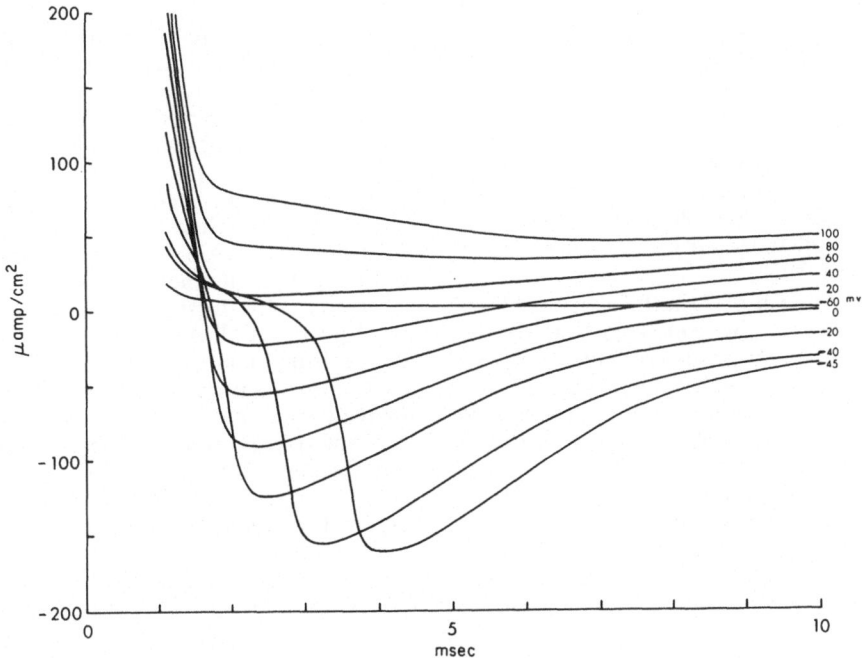

Figure 4. Clamp control current for a series of depolarizing steps in the model of Figure 3*B* (reproduced from 10).

in the sucrose gap and a lumped resistance in series with all of the membrane. The presence of these resistances was well known and typical values were extracted from the literature. Figure 4 shows a set of clamp currents computed for a series of repolarizing steps. Looking at the voltage and current as usually recorded in a single sucrose gap experiment, there appeared to be a fast potential clamp and good (i.e., familiar) current records.

We then asked, as one can in a simulation: Is the potential that we are measuring really the transmembrane potential and is it uniformly controlled? The answer was a resounding "NO"! From the distributions of potential and current in time and space as shown in Fig. 5, it can be

seen that what resulted from a clamp step was in fact a propagating action potential. We were able to show that, for these early currents, the peak current-voltage relationship, which was quite normal in appearance, was virtually independent of membrane characteristics and that it measured only the resistance in series with the membrane and the amplitude of the action potential.

While the preceding simulation did not prove that the early currents could never be studied in cardiac cells, we concluded that it would be wiser to try first to decipher the currents responsible for the repolarization phase of the cardiac action potential and postpone analysis of the earlier currents. There was no shortage of theories or experimental data on "slow currents," but the problem was to decide which results were ambiguous because of deficiencies in the clamp method and which (if any) were not. We decided that we would use the simplest membrane models possible consistent with such unambiguous results, adding new currents or other complications only when forced to do so by new unequivocal data.

We assumed (in the absence of evidence to the contrary) that the instantaneous ionic current-voltage relationships were linear, i.e., that the ionic characteristics of the membrane could be described in terms of one or more conductances g. Next, the functional dependence of these conductances had to be decided. Hodgkin and Huxley (6) found that in the squid axon the membrane conductances were a function of the transmem-

brane voltage V_m and time t (i.e., $g = g(V_m,t)$) as Noble (13) later postulated for the cardiac membrane. Brady and Woodbury (2), on the other hand, suggested (as did Johnson and Tille (8)) that during the repolarization phase the conductances might depend on time only ($g = g(t)$), whereas Kootsey and Johnson (11) proposed exactly the opposite: that the membrane may be in steady state during repolarization ($g = g(V)$). Of all the assumptions as to the functional dependency of the membrane conductances, the latter proved the most attractive. The current-voltage relationship for the membrane (during repolarization) could be computed directly from the action potential, the result being an n-shaped curve with a negative slope region, but no inward current above the resting potential. Furthermore,

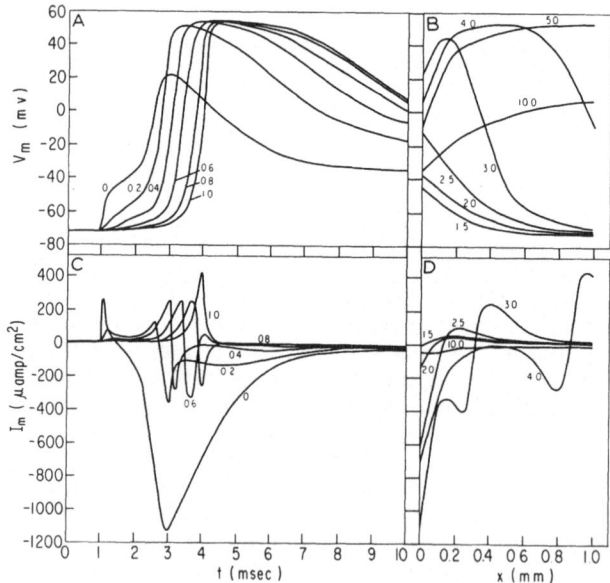

Figure 5. Transmembrane potential V_m and current I_m along the "voltage-clamped" cable (Fig. 3B) as a function of time and distance from the current injection point. Clamp step from resting potential (-72 mV) to -40 mV applied at $t = 1$ ms. A) $V_m(t)$ at six different positions (in mm) along the cable. B) $V_m(x)$ at seven times (in ms) after the clamp step. C) $I_m(t)$ at the same positions as in A. D) $I_m(x)$ at the same times as in B. (reproduced from 10.)

we found that models of this type ($g(V)$) are consistent with the results of many kinds of measurements on cardiac cells including a high input slope resistance during the plateau.

The one experiment that seemed to be clearly contrary to a $g(V)$ membrane was a voltage-clamp experiment on sheep Purkinje bundles made by Weidmann (18) and subsequently by Vassale (17). Their experiment consisted of the initiation of an action potential followed by a momentary polarizing clamp to a fixed potential during the plateau. After 20 to 100 ms the clamp circuit was opened and the membrane potential, rather than moving towards the resting potential, frequently moved in the depolarizing direction, implying that the membrane current-voltage relationship gave an inward current at the clamp potential. They also observed that this phenomenon had an all-or-nothing characteristic, i.e., for very small polarizing clamp steps, the potential returned towards the time course of an unmodified action potential, repolarization after the clamp step continuing as though it had not been disturbed. As the clamp pulse was made more and more negative, there was a sudden and abrupt decrease in the duration of the action potential, the potential returning more rapidly to the resting value at the end of the clamp step.

It was clear that the $g(V)$ membrane would not show either of these characteristics—net inward current during repolarization or all-or-nothing repolarization. We wondered if a preparation would show these characteristics if the $g(V)$ membrane were put into distributed form. Figure 6 shows the results from a simulation of a 6 mm preparation momentarily clamped at one end. Note that the input voltage does show the temporary depolarization after the clamp step even though the membrane description contains no inward current at this voltage. The reason for this is that the clamp step polarized only a part of the membrane in the cable; after the clamp was released, this portion was depolarized by current from membrane that had escaped polarization and remained at a higher potential.

The distributed $g(V)$ model did not, however, show the all-or-nothing behavior. We speculated that this might be due to a two- or three-dimensional effect, because it is well known that the potential falls off from a point source more rapidly as the dimensionality increases. We have not yet simulated the two- or three-dimensional case, but we have repeated the experiment on a synthetic strand of cardiac muscle, grown in tissue culture, which we know to have one-dimensional properties. We observed the same continuous gradation of early repolarization with the size of the clamp step as was observed in the simulation. As a consequence, we have concluded that it is not necessary to postulate a net inward current during the repolarization phase of the cardiac action potential and that the $g(V)$ membrane model is not disproved, at least for this particular preparation, by experiments of the kind described by Weidmann and Vassale.

One way to verify whether the membrane conductances are time-independent during repolarization would be to measure the current-voltage (IV) characteristic curve of the membrane by a voltage clamp experiment, see if it varied with time after the early currents were over, and compare it with the IV curve derived directly from the action potential. To avoid the ambiguities resulting from a complex preparation, we decided again to use the synthetic (one-dimensional) strand. The three microelectrode clamp method of Adrian, Chandler, and Hodgkin (1) was selected because it is ideally suited for cable-like preparations.

We decided that the experiment should be simulated in steady state first to determine how short the clamped part of the cable would have to be to insure that the measured *IV* curve would be essentially that of the membrane. The simulation of a steady-state voltage clamp of a cable is rather different from the simulation of the dynamic experiment. It is much simpler because removing the time variable from the partial differential equation changes it into a second order ordinary differential equation. On the other hand, the simulation is more complex because the input current necessary to maintain a desired membrane voltage is

not known (i.e., it is a two-point boundary value problem). Referring to the diagram of the experiment in Fig. 7, the computations went as follows. For each desired voltage at the point V_1, a guess was made as to the value of the axial current i_a at that point. The steady-state cable equation was then integrated to the (left) end of the cable a distance l away. If the resulting current i_a at the end was not zero, a correction was made to i_a at position V_1 and the integration to the end repeated until i_a at the end was sufficiently close to zero. Starting with the last value of i_a at V_1, the cable equation was then integrated a distance l to the right to point V_2 so that the "measured membrane current" (proportional to $V_2 - V_1$) could be calculated.

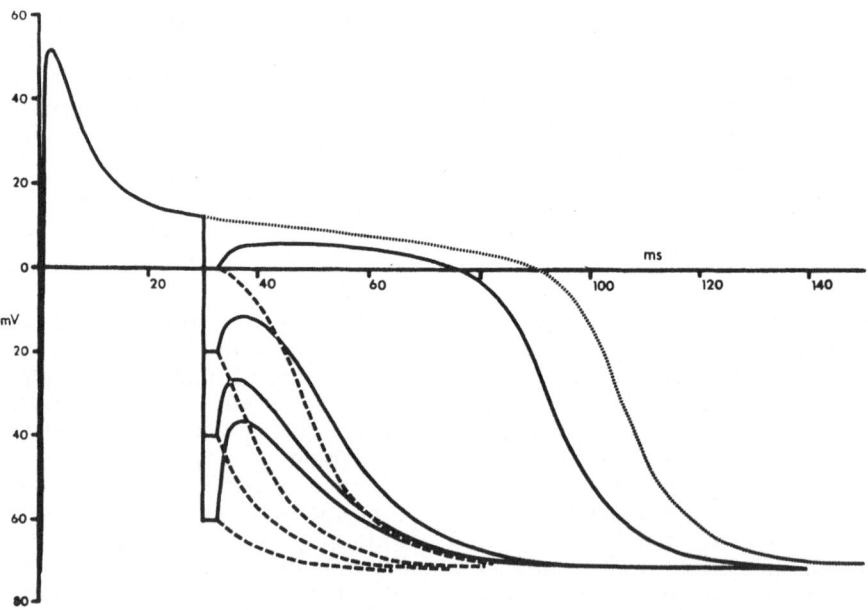

Figure 6. Results of simulation of momentary voltage clamp of the end of a 6 mm cable of 10 μm diameter fibers during the plateau of an action potential. Solid curves show the membrane voltage at the clamped end. Dotted line: undisturbed action potential. Dashed lines: results of similar clamp of single isopotential patch of membrane.

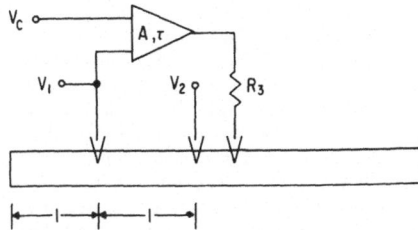

Figure 7. Diagram of the three-microelectrode voltage clamp method (1). V_1 and V_2 are voltage measuring electrodes and current from the clamp amplifier is injected into a third electrode through a series resistance R_3.

The results for three values of l are shown in Fig. 8. When l was very much smaller than the resting length constant λ, the measured IV curve was indistinguishable from that of the membrane. The difference was small but noticeable for $l/\lambda = 0.86$ and the negative slope region disappeared completely for $l/\lambda = 1.72$. Thus the criterion for l in the linear case computed by Adrian, Chandler, and Hodgkin ($l < 2\ \lambda$) is not stringent enough for the nonlinear case.

We have also begun to simulate the dynamic three-microelectrode clamp of the $g(V)$ cable, but so far nothing surprising has emerged except that the clamp is difficult to stabilize and that there are two regions of stability (14).

CONCLUSIONS

It would be nice to be able to conclude by setting out some criteria that could be applied to the results of voltage clamp experiments, criteria that would tell whether an experiment is adequately close to an ideal clamp or not. However, if there is one result that stands out from all of the voltage clamp simulations, it is this:

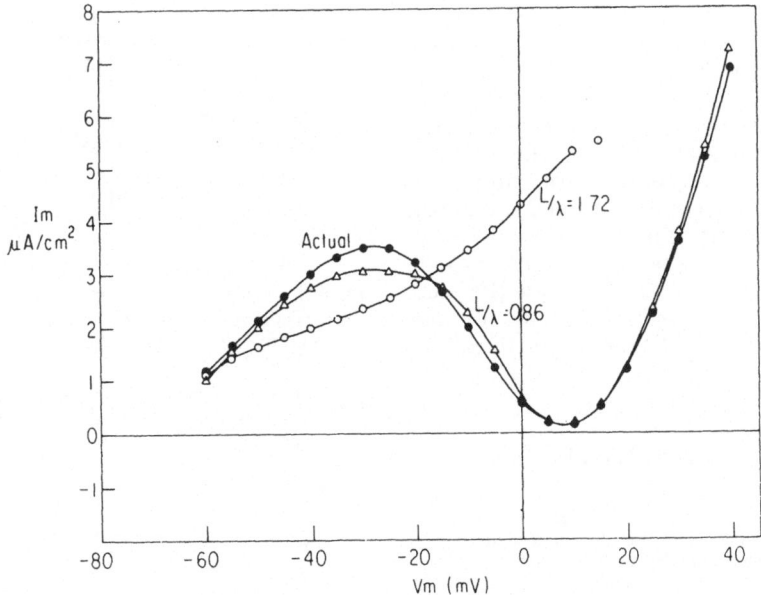

Figure 8. Results of simulation of a steady-state three-microelectrode voltage clamp for two different electrode spacings compared with the actual (assumed) membrane characteristic (filled circles).

It is impossible to tell from the usual records of potential or current versus time alone whether what is recorded is characteristic of the membrane or an artifact of the method. It is unfortunately true that there have been and probably will continue to be published many theories based on experimental results that are really due to the geometry of the preparation and the characteristics of the apparatus.

One might argue that some of these results could have been predicted without computer simulation, particularly those that seem obvious after the fact. However, the human mind simply does not seem capable of predicting in advance all of the consequences of a system of nonlinear partial differential equations. What is needed, clearly, is more insight into membrane potential behavior in space and time. The simulation process is one of the most efficient means of increasing that insight because it literally allows one to do experiments on one's concept of what is going on, while at the same time giving access to the internal variables (potentials, currents, and conductances) that cannot be measured experimentally . Thus, when experimental interventions or simplifications have come to an end (and even when one wants to understand those interventions better) it is still possible to gain information about the electrical properties of cell membranes through realistic simulation of voltage clamp experiments.

REFERENCES

1. ADRIAN, R. H., W. K. CHANDLER AND A. L. HODGKIN. Voltage clamp experiments in striated muscle fibers. *J. Physiol. London* 208: 607, 1970.

2. BRADY, A. J., AND J. W. WOODBURY. The sodium-potassium hypothesis as the basis of electrical activity in frog venticle. *J. Physiol. London* 154: 385, 1960.

3. CARNAHAN, B., H. A. LUTHER AND J. O.

WILKES. *Applied Numerical Methods.* New York: Wiley, 1969.

4. COLE, K. S. Ions, potentials and the nerve impulse. In: *Electrochemistry in Biology and Medicine,* edited by T. Shedlovsky. New York: Wiley, 1955, p. 121–140.

5. GERALD, C. F. *Applied Numerical Analysis.* Reading, Mass.: Addison-Wesley, 1970.

6. HODGKIN, A. L., AND A. F. HUXLEY. A quantitative description of membrane current and its application to conduction and excitation in nerve. *J. Physiol. London* 117: 500, 1952.

7. JOHNSON, E. A., AND M. LIEBERMAN. Heart: excitation and contraction. *Ann. Rev. Physiol.* 33: 479, 1971.

8. JOHNSON, E. A., AND J. TILLE. The repolarization phase of the cardiac ventricular action potential: A time-dependent system of membrane conductance. *Biophys. J.* 4: 387, 1964.

9. JOYNER, R. W., J. W. MOORE AND F. RAMON. *Biophys. J.* 15: 37, 1975.

10. KOOTSEY, J. M., AND E. A. JOHNSON. Voltage clamp of cardiac muscle: A theoretical analysis of early currents in the single sucrose gap. *Biophys. J.* 12: 1496, 1972.

11. KOOTSEY, J. M., AND E. A. JOHNSON. The repolarization phase of the cardiac action potential: An essentially time-independent system of conductance changes. *Biophys. Soc. Abstr.* 13: 130a, 1973.

12. MOORE, J. W., F. RAMON AND R. W. JOYNER. *Biophys. J.* 15: 25, 1975.

13. NOBLE, D. A modification of the Hodgkin-Huxley equations applicable to Purkinje fiber action and pace-maker potentials. *J. Physiol. London* 160: 317, 1962.

14. SCHWARTZ, T. L., AND G. M. KATZ. Temporal control of voltage clamped membrane; Theoretical principles. *Biophys. Soc. Abstr.* 13: 164a, 1973.

15. TASAKI, I., AND A. F. BAK. Discrete threshold and repetitive responses in the squid axon under "voltage clamp." *Am. J. Physiol.* 193: 301, 1958.

16. TAYLOR, R. E., J. W. MOORE AND K. S. COLE. Analysis of certain errors in squid axon voltage clamp experiments. *Biophys. J.* 1: 161, 1960.

17. VASSALE, M. Analysis of cardiac pacemaker potential using a "voltage clamp" technique. *Am. J. Physiol.* 210: 1335, 1966.

18. WEIDMANN, S. Electrophysiologie der Herzmuskelfaser. Bern: Huber, p. 81.

Voltage transients in neuronal dendritic trees

JOHN RINZEL

Laboratory of Applied Studies
Division of Computer Research and Technology
National Institutes of Health, Bethesda, Maryland 20014

ABSTRACT

An analytical method is outlined for calculating the passive voltage transient at each point in an extensively branched neuron model for arbitrary current injection at a single branch. The method is based on a convolution formula that employs the transient response function, the voltage response to an instantaneous pulse of current. For branching that satisfies Rall's equivalent cylinder constraint, the response function is determined explicitly. Voltage transients, for a brief current injected at a branch terminal, are evaluated at several locations to illustrate the attenuation and delay characteristics of passive spread. A comparison with the same transient input applied to the soma shows that the ratio of voltage peaks at two different input sites is, in general, not equal to the ratio of the input resistances. Also for a branch terminal input, the fraction of input charge dissipated by various branches in the neuron model is illustrated. These fractions are independent of the input time course. For transient synaptic conductance change at a single branch terminal, a numerical example demonstrates the nonlinear effect of reduced synaptic driving potential. The branch terminal synaptic input is compared with the same synaptic conductance input applied to the soma on the basis of excitatory postsynaptic potential amplitude at the soma and charge delivered to the soma.—RINZEL, J. Voltage transients in neuronal dendritic trees. *Federation Proc.* 34: 1350–1356, 1975.

The passive response of a neuron to distributed synaptic input depends on the spatiotemporal pattern of that input. The contribution from any individual dendritic input event depends on the location and time course of the input. Questions that naturally arise relate to the voltage transient generated at a single dendritic input site, the degree of attenuation it suffers during passive spread to the soma and other branches, and the total charge actually delivered to the soma.

Wilfrid Rall and I have used a mathematical model for an extensively branched dendritic neuron to address such questions. We have ob-

71

tained analytical expressions for the passive voltage transient at any point in the neuron model for an instantaneous pulse of current applied to a single dendritic branch site. These expressions are used to obtain the transient response for an arbitrary current injected at a single site. Here I will illustrate the response at several locations in the neuron model for a particular transient current applied to a dendritic branch terminal. I will also illustrate the fraction of charge dissipated by various regions of the dendritic model for input at a branch terminal. Corresponding to a dendritic input that is represented as a synaptic conductance change is the nonlinear effect of a reduced synaptic driving potential. This effect is demonstrated with a numerical example for a branch terminal synaptic input. The dendritic input is also compared with the same input applied to the soma. The details of the methods and results for our transient analysis can be found in (29). We have also treated some related steady-state problems (27). Analytic solutions were obtained for the passive voltage response to steady current injection at a single branch site. We used these solutions to calculate the input resistance at branch sites along with steady attenuation factors from the branch site to other locations. The applicability of our neuron model to cat spinal motoneurons, in particular, was discussed there (27).

The dendritic neuron model we employ for our transient analysis is based on a theory for passive dendritic integration developed primarily by Rall (20–26). We assume that the passive membrane potential satisfies the linear one-dimensional cable equation in each dendritic branch segment. We also adopt the branching criterion on which Rall (22) based his "equivalent cylinder" concept. The criterion is: at a branch point, the 3/2-power of the diameter of the parent branch is equal to the sum of the daughter diameters each raised to the 3/2-power. In several theoretical studies of dendritic voltage transients, this branching assumption has been used to treat an entire dendritic tree as a single membrane cylinder: for mathematical treatments, see Rall (21, 22, 24–26), Jack and Redman (9, 10), Redman (28), Barnwell and Cerimele (1), and Norman (18); see Pottala, Colburn, and Humphrey (19) for an electronic model; at this colloquium, Dr. Dodge has described a computational model of a motoneuron with axon which employs the equivalent cylinder concept (5 and unpublished observations). Approaches which lump dendritic branches in this way, do not consider inputs delivered to single branch sites as we do here. Spatially localized transient inputs in dendritic trees have also been treated by Rall (23), and Butz and Cowan (personal communication)[1]; also, at this colloquium, Dr. Barrett has discussed calculations of the transient response to individual inputs in a reconstructed experimental motoneuron (2, 3).

NEURON MODEL AND MATHEMATICAL METHODS

Our neuron model is composed of N identical branching dendritic trees that are coupled at a common origin. This origin is identified as the soma. Each tree has M orders of symmetric branching and each branch point is a bifurcation at which the 3/2-branching criterion is satisfied. In Figure 1A we illustrate the neuron model for the case of six trees $(N = 6)$ and two orders of branching $(M = 2)$. While the geometry shown here is highly idealized our theoretical re-

---

[1] See also the recent publication by E. G. Butz and J. D. Cowan, *Biophys. J.* 14: 661, 1974.

A

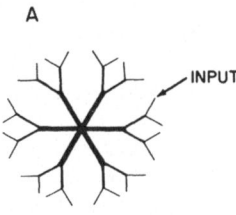

INPUT

B

DENDRITIC
TREE

EQUIVALENT
CYLINDER

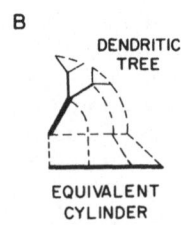

C

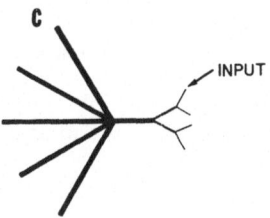

INPUT

Figure 1. Diagrams illustrating features of the idealized neuron model. *A* represents the neuron model composed of six identical dendritic trees with two orders of branching. *B* indicates the relation of a dendritic tree to its equivalent cylinder. *C* represents the neuron model of *A* with dendritic branching shown explicitly only for the tree that receives current injected at the terminal of one branch; the other five trees are represented by their equivalent cylinders. In diagrams *A* and *C* the point of common origin of the trees is regarded as the neuron soma.

sults do not require such a degree of symmetry; for more general considerations I refer to (27, 29). The branching assumption implies that a tree that receives input only at its origin can be treated as a single equivalent cylinder (Figure 1*B*). Consequently, when input is delivered to one branch site the branching details of only the input tree must be

considered; Figure 1*C* illustrates this idea.

The cable properties of a dendritic tree can be described in terms of electrotonic distance (22). In any branch segment of diameter d the electrotonic length constant λ is defined by $\lambda = [(R_m/R_i)(d/4)]^{\frac{1}{2}}$ where R_m is the membrane resistance (ohm cm^2) and R_i is the resistivity (ohm cm) of the intracellular medium. The electrotonic distance X from the origin (soma) to any point x in the tree is the sum of the physical distance increments in each branch segment along the way, each divided by the appropriate length constant for that segment; i.e.,

$$X = \int_0^x [1/\lambda(y)]dy \qquad (1)$$

If all terminals of a tree are electrotonically equidistant from the soma we denote the length of the tree by L. For cat spinal motoneurons, 1.5 is a typical value for L (6, 16). The input resistance of a tree that satisfies the equivalent cylinder constraint and whose terminals are sealed can be expressed as $R_{T\infty} \coth (L)$. Here $R_{T\infty}$ is the input resistance of a semi-infinite cylinder whose diameter is equal to that of the dendritic trunk; it is defined by

$$R_{T\infty} = (2/\pi)[R_m R_i/d^3]^{\frac{1}{2}} \qquad (2)$$

The input resistance R_N seen at the soma of our neuron model is thus

$$R_N = R_{T\infty} \coth (L)/N$$

Now we formulate our basic mathematical problem for the neuron model: determine the transient distribution of membrane potential when current is injected at a single branch. Let us consider the case of input applied to a single branch terminal; we have obtained results for other input locations as well (29). We suppose the dendritic membrane behaves passively. Thus the transient membrane potential $V(X,T)$, expressed as the deviation from rest, satisfies the one-dimensional cable equation in each branch segment:

$$\frac{\partial^2 V}{\partial X^2} = \frac{\partial V}{\partial T} + V \qquad (3)$$

Here X is electrotonic distance; it is dimensionless. Dimensionless time T is the physical time t divided by τ the membrane time constant; $\tau = R_m C_m$, where C_m is the membrane capacitance (farad cm^{-2}). A typical value for τ in cat motoneurons is 5 msec (8, 16). The branch terminals of the neuron are assumed to be sealed. Hence, except at an input terminal, we have the boundary condition

$$\frac{\partial V}{\partial X} = 0 \quad \text{at } X = L \qquad (4)$$

At the input terminal, the axial current in the branch must equal the injected current $I(T)$. This boundary condition can be expressed as

$$\frac{\partial V}{\partial X} = 2^M R_{T\infty} I(T) \quad \text{at } X = L \qquad (5)$$

In addition to these boundary conditions at the terminals, a solution satisfies certain matching conditions at the branch points and at the origin: membrane potential is continuous and core current is conserved (Kirchoff's law). To complete the formulation we specify that the membrane is initially at rest, i.e.,

$$V(X,0) = 0. \qquad (6)$$

The solution for the special input current $I(T) = \delta(T)$, where $\delta(T)$ is the Dirac delta function, is the response for an instantaneous point charge placed at the terminal at $T = 0$. This solution is referred to as the transient response function for input at $X = L$; we denote it by $K(X,T; L)$. Since our mathematical problem is linear, the

solution for arbitrary $I(T)$ can be expressed as the convolution:

$$V(X,T) = \int_0^T K(X, T - s; L)I(s)ds \qquad (7)$$

This formula in general would also involve the decay of an initial distribution of membrane potential but because of *equation 6* this contribution does not appear.

Our method of solving *equations 3–6* is to determine $K(X,T,L)$ analytically and then use *equation 7* to evaluate the solution for particular inputs of interest. To obtain $K(X,T;L)$ we first apply the Laplace transform with respect to time. We next solve the transformed problem by superposition methods and then invert the transform of $K(X,T;L)$. The Laplace transform of the response function is denoted by $\tilde{K}(X,p;L)$ where p is the transform variable. It satisfies the ordinary differential equation

$$\frac{\partial^2 \tilde{K}}{\partial X^2} = (p + 1)\tilde{K} \qquad (8)$$

which is the transform of *equation 3*. The boundary conditions, *equations 4* and *5*, become respectively,

$$\frac{\partial \tilde{K}}{\partial X} = 0 \quad \text{at } X = L \qquad (9)$$

and

$$\frac{\partial \tilde{K}}{\partial X} = 2^M R_{T\infty} \quad \text{at } X = L \qquad (10)$$

Equation 10 follows from *equation 5*, with $I(T) = \delta(T)$, because the Laplace transform of $\delta(T)$ equals one.

Equations 8–10 constitute a time-independent problem in which p appears as a parameter. The particular case of p = 0 corresponds to a steady state problem for *equations 3–5* with $I(T) = 1$ in *equation 5*. The method of superposition which we have used to solve this transformed problem has been presented in detail for the steady-state problem (27).

We will not describe the general method here, but rather display and discuss the solution for a particular case. Consider the problem schematized in the upper right of Fig. 2: a pulse of current is applied to a branch terminal. The neuron model has $N = 6$ and $M = 3$; each branch length ΔX is assumed equal to $L/4$. The transformed response function evaluated at the input terminal is given by

$$\frac{\bar{K}(L,p; L)}{R_{T\infty}}$$

$$= \frac{1}{6}\frac{\coth(qL)}{q} + \frac{5}{6}\frac{\tanh(qL)}{q} + \frac{\tanh(q3\Delta X)}{q}$$

$$+2\frac{\tanh(q2\Delta X)}{q} + 4\frac{\tanh(q\Delta X)}{q} \quad (11)$$

where $q = \sqrt{p+1}$. The response function evaluated at other locations is given by equations 17, 18 and 19 of (29). Each term in the above sum corresponds to the solution of a simple component problem. A term like $R_{T\infty}\coth(qL)/q$ corresponds to the response at the terminal end of a cylinder of length L when a pulse of current is applied to that end; the ends of the cylinder are sealed. A term like $R_{T\infty}\tanh(qL)/q$ has a similar correspondence but in this case the origin of the cylinder is clamped to the resting potential. The last three terms in *equation 11* correspond to the branching details in the input tree. For example, the last term represents equalization of potential between the input branch and its sister branch. The inverse transforms of the component terms in *equation 11* have two different time domain representations. Both representations are infinite series, however one converges better for small values of T while the other converges better for large values of T. For example, the inverse transform of $\coth(qL)/q$ has the small T representation

$$\frac{e^{-T}}{\sqrt{\pi T}}\{1 + 2\exp[-(2L)^2/4T]$$

$$+ 2\exp[-(3L)^2/4T] + \cdots\} \quad (12)$$

and the large T representation

$$\frac{e^{-T}}{L}\{1 + 2\exp[-(\pi/L)^2T]$$

$$+2\exp[-(2\pi/L)^2T] + \cdots\} \quad (13)$$

The time domain expressions for the general component solutions are given by equations 21, 22, 24, and 25 in (29). Both of these representations are useful for obtaining biophysical insight from these mathematical solutions. They can also be used to advantage in evaluating solutions with *equation 7*.

DENDRITIC VOLTAGE TRANSIENTS FOR CURRENT INJECTION TO ONE BRANCH

In this section we apply the results described above to evaluate voltage transients for current injection at a single location. In the first two examples, input is applied to one branch terminal of the neuron model displayed in Figs. 2 and 3. For the last example, input is applied to the soma. The neuron parameters are $N = 6$, $M = 3$, $L = 1.0$, and $\Delta X = 0.25$ (branch length).

Pulse of current (point charge)

First we consider the voltage response to a pulse of current, i.e., an instantaneous point charge, applied to a branch terminal. The voltage transient $K(L,T; L)$ at the input terminal is given by the inverse transform of *expression 11*. This transient is illustrated as the solid curve in Fig. 2; the ordinate scale here is logarithmic. The dashed curves in Fig. 2 correspond to the limiting behavior of the response function as T approaches zero and as T becomes large. We have obtained analytic expressions for these limiting behaviors by using the small T and large T representations of the response function (29). The upper dashed curve is proportional to the response at the origin of a semi-infinite cylinder when a point charge is placed there. The lower dashed curve, a straight line, is proportional to the decay of a spatially uniform potential distribution in a one dimensional cable structure with sealed ends. Intuitively, these asymptotic results can be interpreted as follows. For early time, before a significant amount of charge

has spread beyond the input branch, the voltage at the input terminal responds as if the input branch were a semi-infinite cylinder. After a sufficiently long time however, charge becomes nearly uniformly distributed over the whole neuron and the voltage decays like $\exp(-T)$.

Brief transient current

Now we apply, to a branch terminal, the transient current illustrated in the upper left of Fig. 3. This input has a brief time course given by

$$I(T) = I_p 50T \exp(1-50T) \qquad (14)$$

where I_p is the peak value of $I(T)$.

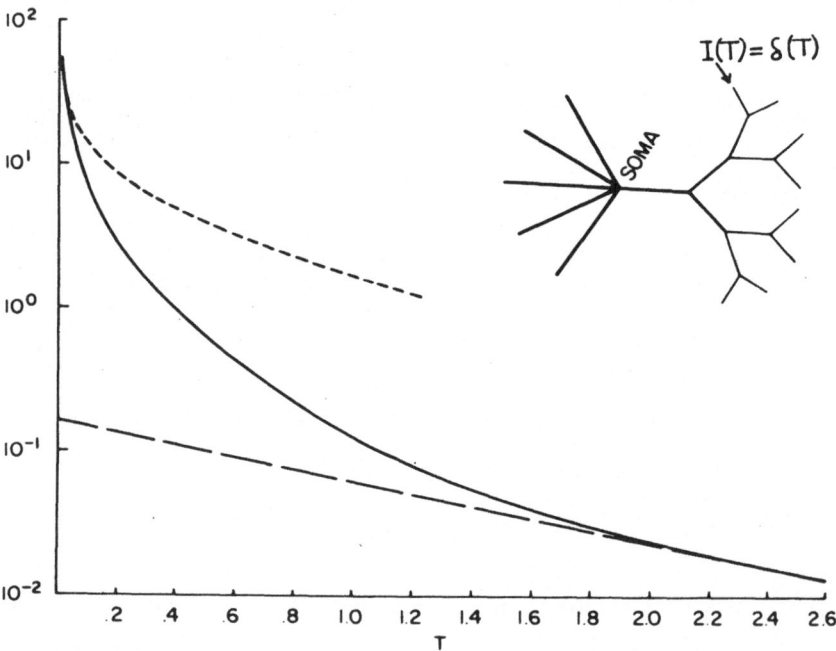

Figure 2. Response function at the input branch terminal (shown solid), compared with two limiting cases (shown dashed); the ordinate scale is logarithmic. The solid curve represents values of $K(L,T;L)/R_{T\infty}$, given by the inverse Laplace transform of *equation 11*. The neuron model, shown upper right, has the parameter values: $M = 3$, $N = 6$, $L = 1.0$, and $\Delta X = 0.25$ (branch length). The lower dashed curve represents the limiting behavior of the response function as $T\rightarrow\infty$; it corresponds to uniform potential decay. The upper dashed curve represents the limiting behavior of the response function as $T\rightarrow0$; it corresponds to the response for a point charge placed at the end of a semi-infinite length of terminal branch.

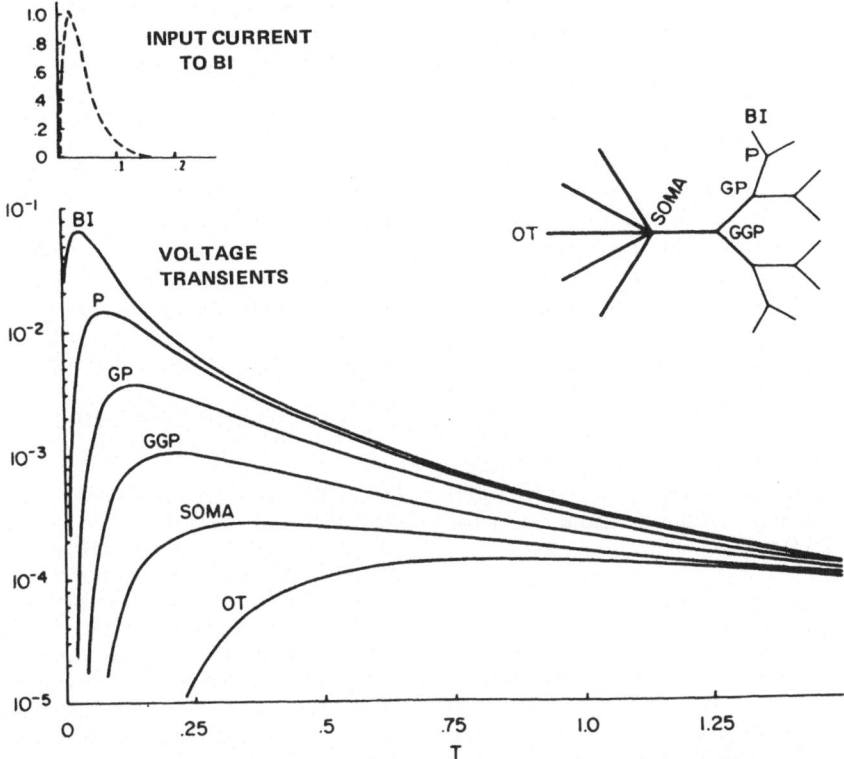

Figure 3. Semi-log plots of transient membrane potential versus T at several locations along the mainline in the neuron model for transient current injected into the terminal of one branch. The input current is given by *equation 14* and shown here as $I(T)/I_p$. BI designates the input branch terminal while P, GP, and GGP designate the parent, grandparent, and great grandparent nodes, respectively, along the mainline from BI to the soma. The response at the terminals of the terminals of the trees not receiving input directly is labeled OT. The neuron model parameters are the same as in Fig. 2. The ordinate scale represents dimensionless values of $V(X,T)/(2^M R_{T\infty} I_p e)$ where $V(X,T)$ is obtained by using the convolution formula, *equation 7*.

The peak value is attained at $T = 1/50$; for example, if $\tau = 5$ msec, the time of peak would be 100 μsec. Inputs with this time course have previously been used in theoretical calculations of dendritic voltage transients (2, 3, 7, 9, 10, 28). To determine the voltage transients at various locations in the neuron model we use the transient response function $K(X,T;L)$ in *formula 7*. Our calculated voltage transients are illustrated in the lower left of Fig. 3; the curves are labeled in accordance with the schematic diagram on the right. They correspond to the input branch terminal(BI), the parent node(P), the grandparent node(GP), the great grandparent node(GGP), the soma and the terminals of the five other trees(OT) that do not receive input directly. These illustrative transients clearly exhibit the smearing and dissipative characteristics of passive spread. The time course of the depolarization is delayed and broadened while the peak amplitude is attenuated at successively more distant locations from the input terminal. In particular, we compare the transients at the input site and at the soma: the peak times are 0.04 and 0.35; the half-widths are 0.09 and 1.0; and the peak amplitudes are 64.8×10^{-3} and 0.276×10^{-3}. The ratio of these peak amplitudes, the transient attenuation factor, equals about 235 in this case. It should be noted that this factor, as well as the illustrative transients, depends on the input time course. For a very slow input the transient attenuation factor will be nearly equal to the steady-state attenuation factor of 23.9 (27).

Comparison for input to different locations

Since the scale for the preceding example is arbitrary it is useful to have a reference case for comparison. For this purpose we apply the current, given by *equation 14*, to the soma ($X = 0$) of our neuron model. The transient voltage distribution is calculated by using *equation 7* with $K(X,T; L)$ replaced by the response function for input at the soma $K(X,T; 0)$. The peak amplitude of the voltage transient at the soma is 1.42 $\times 10^{-3}$ (relative to the ordinate scale in Fig. 3). On comparison with the preceding example of a branch terminal input, the ratio of voltage peaks at the two input locations is 46.3 for this $I(T)$. Hence if the soma peak is 1 mV for this $I(T)$ applied to the soma, then when the same input is applied only to a branch terminal the peak at the terminal would be 46.3 mV. Although this comparison is for a specified time course of injected current, it suggests that the nonlinear effects associated

with brief synaptic conductance input could be relatively significant for the branch location as compared to the soma. This matter is considered in a later section. The comparison also reveals that the ratio of voltage peaks at the two input sites, for the given brief current, is not equal to the ratio of their input resistances; here, the input resistance ratio equals 15.6 (27). Although there are certain applications in which these ratios are equal, or have been assumed so (11, 12), in general they are not equal (28, 29).

CHARGE DISSIPATION IN NEURON MODEL

For a given transient current injection there is a certain amount of charge delivered to the neuron. As the potential distribution runs its time course, this input charge redistributes itself and leaks outward across the membrane of the various dendritic branches. Estimates for the fraction of input charge that is dissipated by the soma can be useful in comparing inputs at different locations. We have obtained the distribution of charge dissipation over our neuron model for transient input to a single location. Questions related to charge dissipation have also been considered by Barrett and Crill(3), Iansek and Redman(7), and Redman(28).

Under the assumptions of one-dimensional cable theory the core current, i_{core}, through a cross section of a k-th order dendritic branch satisfies

$$2^k R_{T\infty} i_{core} = -\frac{\partial V}{\partial X} \qquad (15)$$

The total charge dissipated by the membrane resistance in the segment of this branch between X_a and X_b is denoted by $Q(X_a, X_b)$. By definition, it is the difference between the charge which entered at X_a (the time integral of the core current at X_a) and that which left at X_b during the entire

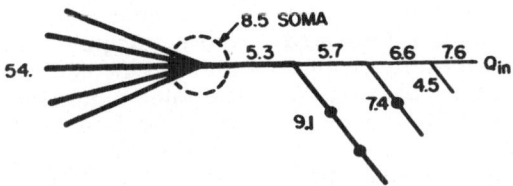

Figure 4. Percentages of total input charge Q_{in} dissipated in different regions of the neuron model for transient current injection at a single branch terminal. The model neuron parameters and branching diagram are the same as in Fig. 2. The various side paths that leave the mainline from the input site to the origin are represented by their respective equivalent cylinders. A soma that corresponds to the segment $X = 0$ to $X = 0.1$ of each dendritic trunk dissipates 8.5% of Q_{in}. The percentages shown here were calculated by using *equations 15* and *16* and the steady-state solutions of Rall and Rinzel (27).

transient event; it is given by

$$Q(X_a, X_b) = \int_0^\infty i_{core}(X_a)dt$$

$$- \int_0^\infty i_{core}(X_b)dt \quad (16)$$

If we use the *relationship 15* in *equation 16*, we see that $Q(X_a, X_b)$ can be expressed in terms of derivatives with respect to X of the time integral of $V(X,T)$. This time integral, however, satisfies a steady state problem (1, 28, 29) similar to *equations 8–10* with p = 0. Consequently, the charge dissipation in any segment can theoretically be determined without actually integrating the voltage transients explicitly. We have used this observation along with our steady-state solutions of (27) to calculate the percentage of input charge Q_{in} dissipated in the various dendritic branches of our neuron model for input to a single branch terminal. Our results are illustrated in Fig. 4 for the neuron model displayed in Figs. 2 and 3. Here we have gathered together various side branches off the mainline from the input terminal to the soma. Summarizing the calculations we find: about 25% of Q_{in} is dissipated along the main line; the side branches off the mainline account for 21% of Q_{in}; while the other five trees dissipate 54% of Q_{in}. Sup-

pose we consider the soma to be composed of an initial electrotonic length increment of 0.1 from each dendritic trunk. We then find that 8.5% of the Q_{in} applied at a branch terminal is dissipated by the soma.

For comparison, we have made a similar calculation for input applied to the soma. We find that 12.7% of Q_{in} applied at $X = 0$ is actually dissipated by the soma. It should be remembered however that such comparisons take no account of the transient characteristics of the input; the charge dissipation percentages are independent of the input time course. For applications in which temporal aspects of inputs are important one must consider the voltage transients themselves. We remark that these charge dissipation results also apply to periodic inputs. In that case, Q_{in} would be the charge delivered per period and the percentages shown in Fig. 4 would correspond to the charge dissipated over a single period.

SYNAPTIC INPUT AS A CONDUCTANCE CHANGE AT ONE LOCATION

The results described in the previous sections have considered input to the neuron model in the form of a specified current injection. However when synaptic input is represented as a membrane conductance change, the synaptic current depends on the excitatory postsynaptic potential (EPSP) at the input site; e.g., see (23). When the local depolarization is significant compared to the synaptic equilibrium potential there can be a sizable reduction in the effective synaptic driving potential. For a given conductance change, the degree of this reduction will depend on the input location. We have used our transient results to illustrate these effects, often referred to as nonlinear synaptic effects, with a computational example. Calculations that consider transient synaptic nonlinearities have also been performed by Barrett and Crill (3), Rall (23, 24), and Mac-Gregor (17).

Let $g_\epsilon(T)$ be the transient excitatory conductance for a synaptic input at X_{in}. The synaptic current $I_\epsilon(T)$ is defined by

$$I_\epsilon(T) = g_\epsilon(T)[V_\epsilon - V_{in}(T)] \quad (17)$$

where V_ϵ is the synaptic equilibrium potential (minus the resting potential), assumed constant, and $V_{in}(T)$ is the transient depolarization at X_{in}. Suppose $I_\epsilon(T)$ is known, then the voltage response at any location in the neuron model could be calculated by using a convolution formula like *equation 7*. In particular, the EPSP at the input location X_{in} satisfies

$$V_{in}(T) = \int_0^T K(X_{in}, T - s; X_{in}) I_\epsilon(s) ds \quad (18)$$

where $K(X,T;X_{in})$ is the transient response function for input at X_{in}. If the *expression 17* for $I_\epsilon(T)$ is substituted into *equation 18* we find that $V_{in}(T)$ satisfies an integral equation. Once $V_{in}(T)$ is determined as the solution to *equation 18*, then $I_\epsilon(T)$ is known and the voltage transients at other locations can be determined by convolution.

We have solved numerically the above integral equation for $V_{in}(T)$ with specified $g_\epsilon(T)$ and X_{in}. We used the same neuron model parameters as in Figs. 2–4 and, for this example, the whole neuron input resistance was specified as 1 megohm. For $g_\epsilon(T)$ we used the brief time course given by *equation 14* with I_p replaced by 10^{-7} mho. For the discussion of these calculations in (29) we did not assign particular values to V_ϵ and τ; here, I will set $V_\epsilon = 70$ mV and $\tau = 5$ msec. First, we applied the synaptic conductance change to the soma ($X = 0$) of our neuron model. The resulting EPSP had a peak value of about 1 mV at the soma. This amplitude lies near the upper end of the range of somatic unitary EPSP's recorded from cat spinal motoneurons (4, 7, 8, 13). Because the EPSP amplitude is small relative to V_ϵ the reduction in driving potential is negligible for this case of input to the soma.

Next, we applied the synaptic input $g_\epsilon(T)$ to a single branch terminal ($X_{in} = L$). The solid curves in Fig. 5 illustrate the synaptic current $I_\epsilon(T)$ and the EPSP $V_{in}(T)$ at the input terminal. The peak values of these transients are 4.8 nanoamp and 28.8 mV, respectively. Also shown (dashed) here are the reference transients, I_{ref} and V_{ref}, which were calculated with V_{in} replaced by zero in *equation 17*. The reference case thus assumes that the synaptic driving potential is constant. We use these reference transients to quantitatively illustrate the effects of reduced synaptic driving potential at a single branch location. A comparison of the two sets of curves reveals that peak I_ϵ is 68.2% of peak I_{ref} and peak V_{in} is 64% of peak V_{ref}. Also, the total synaptic input charge (the time integral of I_ϵ) is reduced to 67.2% of the reference value. The slight variation in these three reduction percentages is due to the difference in time course of I_ϵ and I_{ref}. These percentages, as well as their relative differences, are of course less for smaller amplitude conductance changes. For example, with the same time course for $g_\epsilon(T)$ but with a peak amplitude of 10^{-8} mho, we find that the three reduction percentages for the terminal input would all equal 95%. The corresponding EPSP peak at the soma when this input is applied to the soma is 98 μV, which compares to the average amplitude of a quantal EPSP of somatic origin reported by Kuno and Miyahara(14, 15).

Finally we compare, for synaptic input (peak value equal to 10^{-7} mho) applied separately to the two locations, the EPSP peaks at the soma and the amounts of charge dissipated by the soma. After passive spread, the attenuated EPSP amplitude at the soma for the branch terminal input is 129 μV. The ratio of this peak value to the peak value of the soma EPSP for input at the soma is nearly one-seventh. To calculate the charge dissipated by the soma for each of the two inputs, we apply the results of the preceding section. For synaptic input to the soma, 0.239×10^{-12} coulomb is dissipated by the soma. The corresponding figure for the branch terminal synaptic input is 0.109×10^{-12} coulomb. Hence the amount of charge dissipated by the soma for the synaptic input at a branch terminal is nearly one-half of that dissipated by the soma when the same synaptic conductance change occurs at the soma. Barrett and Crill(3) have also made numerical comparisons for synaptic inputs to the soma and to single dendritic branches.

SUMMARY

An extensively branched dendritic neuron model which assumes the

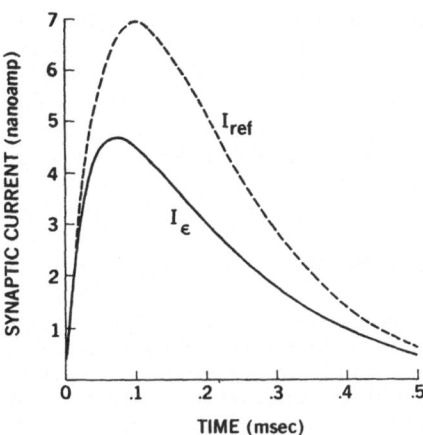

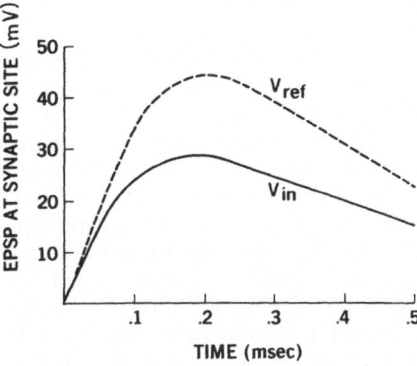

Figure 5. Comparison of current and voltage transients for synaptic conductance change at a branch terminal. The transient synaptic current I_ϵ and depolarization V_{in} at the input terminal (shown solid) are the solutions of *equations 17* and *18*. The reference transients, I_{ref} and V_{ref}, (shown dashed) were calculated assuming constant synaptic driving potential; they also satisfy *equations 17* and *18*, but with $V_{in} = 0$ in *equation 17*. The time course of the synaptic conductance change is given by *equation 14* with I_p replaced by 10^{-7} mho. The neuron model parameters are the same as in Fig. 2; here, $\tau = 5$ msec, $V_\epsilon = 70$ mV, and the whole neuron input resistance equals 1 megohm.

equivalent cylinder constraint on branching is described (Fig. 1). The mathematical problem that describes the passive voltage response for arbitrary current injection at a single branch terminal is formulated. The solution is expressed as a convolution of the input current with the transient response function, the solution for an instantaneous pulse of current. The response function has been determined explicitly by using Laplace transforms and superposition methods; for details, see (27, 29).

For a particular brief input current applied to a single branch terminal, the convolution formula is used to evaluate the voltage transients at several locations in the neuron model. These transients illustrate the attenuation and delay characteristics of passive spread (Fig. 3). The transient at the input branch is compared to the soma transient obtained when the same current is applied to the soma. The ratio of the peak depolarizations is not equal to the ratio of the input resistances as it would be if the input were steady (independent of time).

While the transient voltage distribution depends on the input time course, the fraction of charge dissipated by each branch of the neuron does not. The distribution of charge dissipation is illustrated for input to a branch terminal (Fig. 4); in this example, 8.5% of the input charge to the branch terminal is dissipated by the soma.

For transient synaptic conductance change at a dendritic branch terminal, a numerical example demonstrates the nonlinear effect of reduced synaptic driving potential (Fig. 5). Quantitative comparison with a reference case, which assumes that the synaptic driving potential is constant, shows that the peak synaptic current is reduced to 68% of the peak reference current and the peak EPSP at the input terminal is reduced to 64% of the peak reference voltage.

The branch terminal synaptic input is also compared with the same synaptic conductance change applied to the soma: on the basis of transient EPSP peak at the soma, the branch input is nearly one-seventh as potent as the somatic input; on the basis of charge delivered to the soma, the potency of the branch input is about half that of the soma input.

REFERENCES

1. BARNWELL, G. M., AND B. J. CERIMELE. *Kybernetik* 10: 144, 1972.
2. BARRETT, J. N. *Federation Proc.* 34: 1398, 1975.
3. BARRETT, J. N., AND W. E. CRILL. *J. Physiol. London* 293: 325, 1974.
4. BURKE, R. E. *J. Neurophysiol.* 30: 1114, 1967.
5. DODGE, F. A., AND J. W. COOLEY. *IBM J. Res. Develop.* 17: 219, 1973.
6. IANSEK, R., AND S. J. REDMAN. *J. Physiol. London* 234: 613, 1973.
7. IANSEK, R., AND S. J. REDMAN. *J. Physiol. London* 234: 665, 1973.
8. JACK, J. J. B., S. MILLER, R. PORTER AND S. J. REDMAN, *J. Physiol. London* 215: 353, 1971.
9. JACK, J. J. B., AND S. J. REDMAN. *J. Physiol. London* 215: 283, 1971.
10. JACK, J. J. B., AND S. J. REDMAN. *J. Physiol. London* 215: 321, 1971.
11. KATZ, B., AND R. MILEDI. *J. Physiol. London* 168: 389, 1963.
12. KATZ, B., AND S. THESLEFF. *J. Physiol. London* 137: 267, 1957.
13. KUNO, M. *Physiol. Rev.* 51: 657, 1971.
14. KUNO, M., AND J. T. MIYAHARA. *J. Physiol. London* 201: 465, 1969.
15. KUNO, M., AND J. T. MIYAHARA. *J. Physiol. London* 201: 479, 1969.
16. LUX, H. D., P. SCHUBERT AND G. W. KREUTZBERG. In: *Excitatory Synaptic Mechanisms*, edited by P. Anderson and J. K. S. Jansen. Oslo: Universitetsforlaget, 1970, p. 189.
17. MACGREGOR, R. J. *Biophys. J.* 8: 305, 1968.
18. NORMAN, R. S. *Biophys. J.* 12: 25, 1972.
19. POTTALA, E. W., T. R. COLBURN AND D. R. HUMPHREY. *IEEE Trans. Bio-Med. Eng.* BME-20: 132, 1973.
20. RALL, W. *Exptl. Neurol.* 1: 491, 1959.
21. RALL, W. *Exptl. Neurol.* 2: 503, 1960.
22. RALL, W. *Ann. N.Y. Acad Sci.* 96: 1071, 1962.
23. RALL, W. In: *Neural Theory and Modeling*, edited by R. F. Reiss. Stanford, CA: Stanford Univ. Press, 1964, p. 73.

24. RALL, W. *J. Neurophysiol.* 30: 1138, 1967.

25. RALL, W. *Biophys. J.* 9: 1483, 1969.

26. RALL, W. In: *Excitatory Synaptic Mechanisms*, edited by P. Anderson and J. K. S. Jansen. Oslo: Universitetsforlaget, 1970, p. 175.

27. RALL, W., AND J. RINZEL. *Biophys. J.* 13: 648, 1973.

28. REDMAN, S. J. *J. Physiol. London* 234: 637, 1973.

29. RINZEL, J., AND W. RALL. *Biophys. J.* 14: 759, 1974.

Propagation of action potentials in inhomogeneous axon regions[1]

F. RAMÓN,[2] R. W. JOYNER AND J. W. MOORE

Department of Physiology and Pharmacology
Duke University Medical Center, Durham, North Carolina 27710

ABSTRACT

Described are studies of propagation of action potentials through inhomogeneous axon regions through experiments performed on squid giant axons and by computer simulations. The initial speed of propagation of the action potential is dependent upon the stimulus waveform. For a rectangular pulse of current, the action potential travels initially at a high speed that declines over the distance, reaching a constant speed of propagation at about 1–5 resting length constants; this distance depends on the stimulus strength. Additional experiments studied the effects of changing the axon diameter and of introducing a temperature step. It was found that the propagated action potential suffers profound modifications in shape and velocity as it reaches the region of transition. In both cases, it was possible to obtain reflected action potentials. A region of increased effective diameter was produced experimentally in the squid giant axon by insertion of an axial wire as usually employed in voltage clamps. It was found that the action potential, at the axial wire tip region, undergoes shape changes similar to those obtained in simulations of a region of increased diameter as in a junction with the axon and soma in motor neurons. It is concluded that the giant axon can be used to reproduce simple electrical behaviors in other structures.—RAMÓN, F., R. W. JOYNER AND J. W. MOORE. Propagation of action potentials in inhomogeneous axon regions. *Federation Proc.* 34: 1357–1363, 1975.

The squid giant axon is ideally suited not only to study the properties of an excitable membrane, but also may be used to experimentally reproduce the electrical behavior of some other excitable cells. It is this second approach that motivated this paper. For example inhomogeneities may be introduced into the normal axon to represent, in a simplified way, some of the electrical characteristics associated with alterations in more com-

[1] Supported by National Institutes of Health Grant No. NS 03437.
[2] Grass Fellowship in Neurophysiology.

plex cells such as a changing diameter, and so on.

Digital computer simulations of such experiments are also very useful in giving insight into the basis of the experimentally observed phenomena. In order to understand the propagation of action potentials in axons in which inhomogeneities have been introduced, of course the simpler case of propagation along a homogeneous axon has to be well understood first.

Action potential propagation is described by a partial differential equation and has been solved numerically for a variety of conditions. By transforming the partial differential equation into an ordinary differential equation for wave propagation at a constant speed, Hodgkin and Huxley (10) computed the first propagated action potential. Cooley and Dodge (3) later solved the full partial differential equation for propagation without the constant velocity restriction.

Propagation of action potentials in homogeneous unidimensional cables simulating the axon have been studied rather completely (3, 19, 23, 24). Simulations have been carried out for the introduction of certain inhomogeneities into restricted portions of the cable. Examples of these inhomogeneities are: a) cable regions of abnormal external concentration of ions (6); b) cable regions affected by pharmacological agents or toxins (6); c) cable regions of different temperature (11); and d) regions of increased diameter (13, 16). A much more complete case of an inhomogeneous cable has been the simulation of a motor neuron (7).

In this paper we would like to present first some examples (computations) of action potential propagation for homogeneous as well as inhomogeneous cables. Lastly we will show that use of the squid giant axon can be extended to a biological model of another excitable structure, the soma cell. This condition was obtained experimentally by insertion of an axial wire as frequently used to obtain spatial uniformity for voltage clamp purposes[3].

METHODS

Computations

The computations, except where stated otherwise, were obtained on a cable with the following characteristics: diameter 500 μm and length 5 cm, divided in 100 equal segments of 500 μm each. All other parameters were the same as in the original Hodgkin and Huxley model (10).

The partial differential equations were solved numerically by implicit integration methods (5)[4] on a PDP-8/E digital computer with a Hardware Floating Point Processor, and plotted on a Tektronix 611 Storage Oscilloscope. Permanent records were made from the screen of the oscilloscope by a Tektronix Hard Copy Unit.

Experimental

The experiments were performed on giant axons obtained from squids supplied by the Marine Biological Laboratory at Woods Hole, Massachusetts. The solution in the bath was normal artificial sea water at a temperature of about 15 C.

The experimental setup is diagrammed in Fig. 1. The axial wire (75 μm in diameter) was platinized and its resistance measured according to currently used methods (17). Intracellular voltages were recorded through an axial wire (25 μm in diameter) completely isolated except at its tip,

---

[3] A brief account of some of the findings included in this paper has been presented (21).

[4] A general description of the numerical methods is given in the preceding paper by Kootsey (*Federation Proc.* 34: 1343, 1975) and a detailed description of our method has been reported (18).

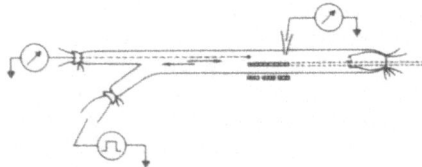

Figure 1. Schematic diagram of the experimental setup used to approximate, on squid giant axons, a region of increased diameter. An axial wire was introduced to the axon from the right side. This was the regular axial wire used for voltage clamp purposes. The axon was dissected up to its branches. One of the branches was used to stimulate an action potential by means of a suction electrode. Through the other branch a wire, isolated except at its tip, was introduced to record the intracellular potential. Occasionally, a microelectrode (5 MΩ resistance) was also used for intracellular recording.

or with microelectrodes of about 5 MΩ resistance. Permanent records were obtained from the screen of a dual beam oscilloscope (type 565, Tektronix, Inc.) and photographed with an oscillographic camera (Model C4, Grass Instrument Co.).

RESULTS

In general, the numerical solution for the voltage profile of a cylindrical cable is obtained by solving the uni-dimensional cable equations derived from the circuit of Fig. 2 (11). It has been frequently pointed out (15, 20, 22) that the core conductor model does not strictly apply to the case of an axon in a large volume of electrolyte because there are external radial voltage gradients as well as longitudinal. Nevertheless this does not significantly affect the accuracy of the transmembrane potential calculations for the core conductor model. Therefore, for convenience, the external resistances are frequently set to zero for axons in a large chamber. Because some of our experiments were conducted in small chambers and we have occasionally used external electrodes to start the action potential, but mainly out of curiosity, we solved the differential equations that describe the circuit diagrammed in Fig. 2 with external resistances.

Homogeneous cable with external resistances

We will consider the situation in which the axon is placed in a cylindrical container about twice its diameter and stimulated with bipolar extracellular electrodes. Figure 3 shows the computed results using an external resistivity equal to that of sea water. A total of 80 segments, each

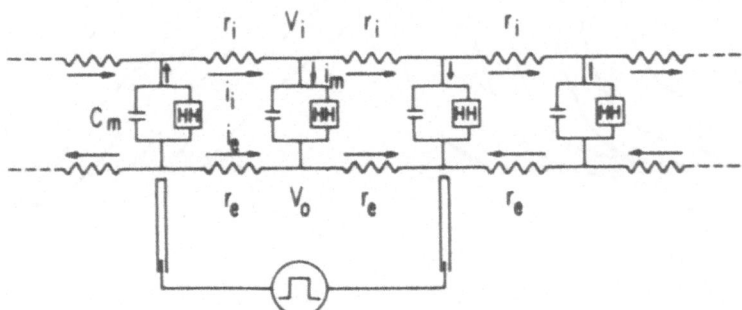

Figure 2. Core conductor model of an axon placed in a chamber. Internal and external resistances and currents are shown. Extra-cellular electrodes, used for stimulation, are also shown.

representing 500 μm, were used for this simulation and the figure shows the transmembrane potential for 6 of them (at 5 mm intervals) as a function of the elapsed time. The two stimulating electrodes were placed adjacent to segment 20 (cathode, *curve 3*) and segment 30 (anode, *curve 4*).

Figure 3 shows, during the first 200 μsec when a stimulating current is applied, the depolarization of the region under the cathode and a symmetrical hyperpolarization under the anode because membrane properties are passive at this time. However, when the stimulus is turned off, the region under the cathode initiates an action potential which propagates in both directions along the cable.

Figure 4 shows the transmembrane potential, as a function of the distance along the cable, at various times after the beginning of the external stimulation. At 0.1 msec depolarization and

hyperpolarization are symmetrical at the locations of the stimulating electrodes. The active membrane response begins at the location of the cathode (*curve b*, 0.310 msec), producing an action potential at that site (*curve d*, 1.065 msec) that propagates in both directions (*curve e*, 1.265 msec).

The propagation of the action potential in the two directions is not symmetrical. Toward the right of the figure the action potential quickly reaches a constant velocity. Toward the left of the figure the action potential propagates into a region that has been previously hyperpolarized and the speed of propagation is slower than the speed to the right (*curve f*, 1.505 msec). After the action potential propagates through the region under the anode, it reaches the same velocity as the action potential propagating to the right.

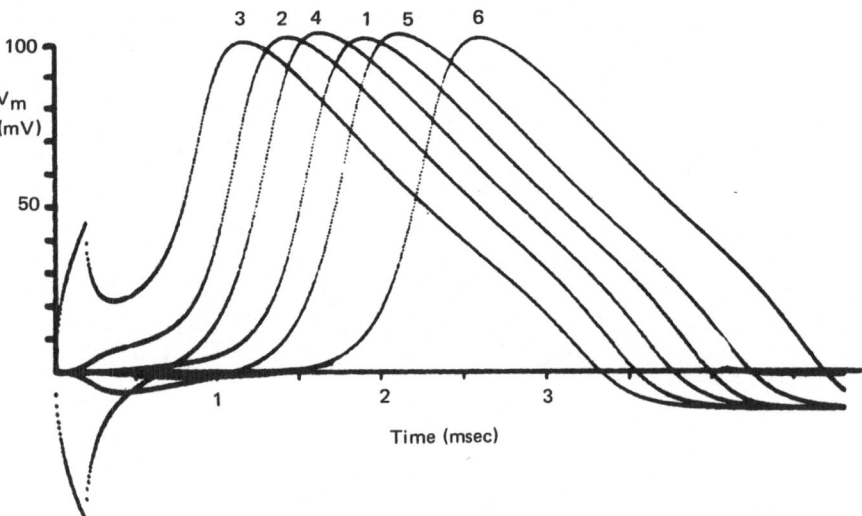

Figure 3. Plot of transmembrane potential as function of time for the equivalent circuit of Fig. 2. The external electrodes were placed facing the segments whose voltages are labeled 3 (cathode) and 4 (anode). Note the initial passive membrane response and, later, the action potential propagating towards the left (*curves 3, 2* and *1*) as well as towards the right (*curves 4, 5* and *6*).

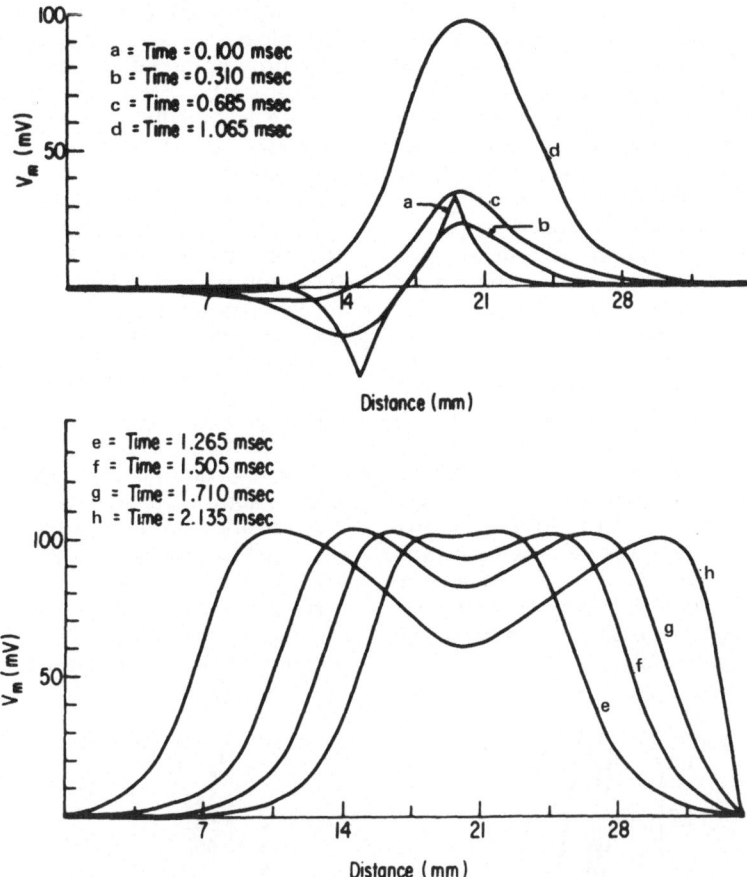

Figure 4. Plot of transmembrane potential as function of the cable distance for the equivalent circuit of Fig. 2. Note at 1.065 msec (*curve d*) the action potential generated at the cathode region. At 1.265 msec the action potential starts propagating toward both ends of the cable. The apparent slow propagation velocity toward the right of the figure is due to the cable termination by a grounded end.

Homogeneous cable without external resistances

Effect of the stimulus waveform

As early as 1938 (9), experimental results established that an action potential travels at a constant speed in nerves. However, action potentials do not obtain a constant speed of propagation until they have moved away from the region where the stimulating current is being injected.

Following the Hodgkin and Huxley (10) idea that the partial differential equation could be converted into an ordinary differential equation for wave propagation, the speed of propagation has been used for the numerical solution of propagated action potentials as a parameter required to obtain convergence of the solution (2, 9). Since in our method for numerical solution of the cable's partial differential equations (see METHODS), the speed of propagation is a result of the computation, not a parameter, we decided to study the way in which the action potential starts traveling and reaches its constant speed of propagation. In our computations we took, as an arbitrary index of the arrival of the impulse in a particular cable region, the time (obtained by a linear interpolation procedure) at which the membrane voltage becomes depolarized by 50 mV (from the resting potential). The instantaneous speed was taken as the segment length divided by the difference between the time of arrival of the impulse at two adjacent segments. We used for stimulation a rectangular pulse of current because this is the usual waveform for stimulation in experimentally elicited action potentials.

For a given stimulus the plot of the normalized speed of propagation of an action potential versus distances along the cable is shown in Fig. 5. The velocity is initially very high and it decreases as the action potential moves away from the point of stimulation, until it finally reaches a constant speed of propagation. If several stimulus strengths are used to elicit the action potential, Fig. 5 shows that the action potential always reaches the same constant speed, but at different distances from the point of stimulation.

In Fig. 5 the distance at which the constant speed is finally reached is longer for stimulus just suprathreshold and shorter for stronger stimulus. This result can be understood by noticing that, with just suprathreshold stimulation, the cable region where current is injected has a long latency for the development of the action potential. During this latency, the de-

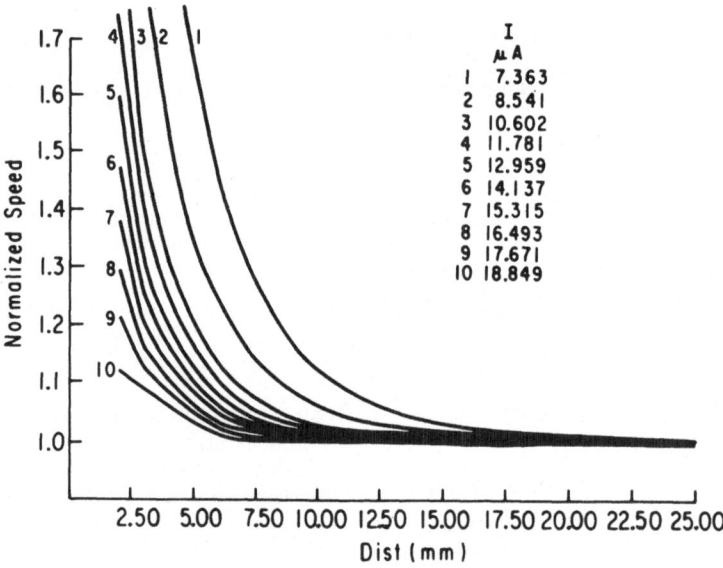

Figure 5. Plots of the normalized speed of propagation of an action potential elicited with square pulse of current of different intensities. Note that, for strong stimuli, the action potential reaches a constant speed of propagation at a shorter distance than for weak stimuli.

polarization of the initial regions spreads along the cable. Therefore, when the action potential finally takes off, it travels on cable regions that are partially depolarized. This depolarization places the membrane potential closer to threshold than normally polarized regions.

The initial acceleration of the action potential depends on the stimulus waveform; in our first case (Fig. 5) this was a rectangular pulse of current. If another stimulus waveform is used, the plot of the initial velocity of the action potential does not have the shape shown in Fig. 5. In the region of a long cable where the constant speed of propagation has been achieved, the normal "stimulus" is the longitudinal intracellular current. When the stimulus at the origin is shaped (with a sum of exponentials) to approximate the longitudinal current waveform (Fig. 6A), the action potential starts traveling at a speed close to the normal constant value. This result is shown in Fig. 6B, where a plot of the velocity obtained with a rectangular current pulse is compared to that obtained by stimulating with the intracellular current waveform.

In both examples of propagation shown before, an action potential elicited by external stimulation (Figs. 3 and 4) and its initial speed after the stimulus (Figs. 5 and 6), the action potential propagated along cable regions whose membrane potential was not at the resting level. To study the effect of membrane polarization on the conduction velocity of an action potential, the conduction velocity is plotted (Fig. 7) as a function of the axon distance from the point of stimulation for normal axons and for axons with a steady 1 or 2 mV depolarization or hyperpolarization. Compared to the control velocity, the conduction velocity is increased by depolarization and decreased by hyperpolarization. The effect is explained by noticing that a depolarized

membrane is closer to threshold than the control. This polarization also slightly affects the shape of the action potential since a depolarization or hyperpolarization causes the membrane to be more or less inactivated than in the control and therefore the rate of rise of the action potential is slower or faster, respectively.

Inhomogeneous conditions

Temperature difference

An example of interaction between two excitable regions can be obtained by assigning, to one restricted portion of the cable, a different temperature than the rest of the cable. The next two Figures (8 and 9) show the results of the propagation in a cable through a region of low temperature, 3 C, bounded by regions of 25 C. Figure 8 shows a normally propagating action potential (curves 1, 2, 3, 4 and 5) changing when it reaches the cold region (curve 6). In the cold region the activation kinetics are slowed down by such a large factor that the action potential persists in that region for a much longer time than in the warm region. While the cold region is still developing an action potential, the region ahead of it (at 25 C) undergoes an action potential that propagates at high speed again (curves 7 and 8).

The action potential in the cold region has such a long duration that the preceding warm region has time to recover its excitability and is capable of being stimulated again to produce another action potential. This second action potential now travels in a reverse direction, toward the point of stimulation (curves 5, 4, 3 and 2 to the right of the figure) The results may be better appreciated in Fig. 9, where they are plotted as functions of the cable distance for various times from the beginning of the stimulation. At 0.670 msec (curve a) the action potential is propagating over the warm region and reaches the cold region at

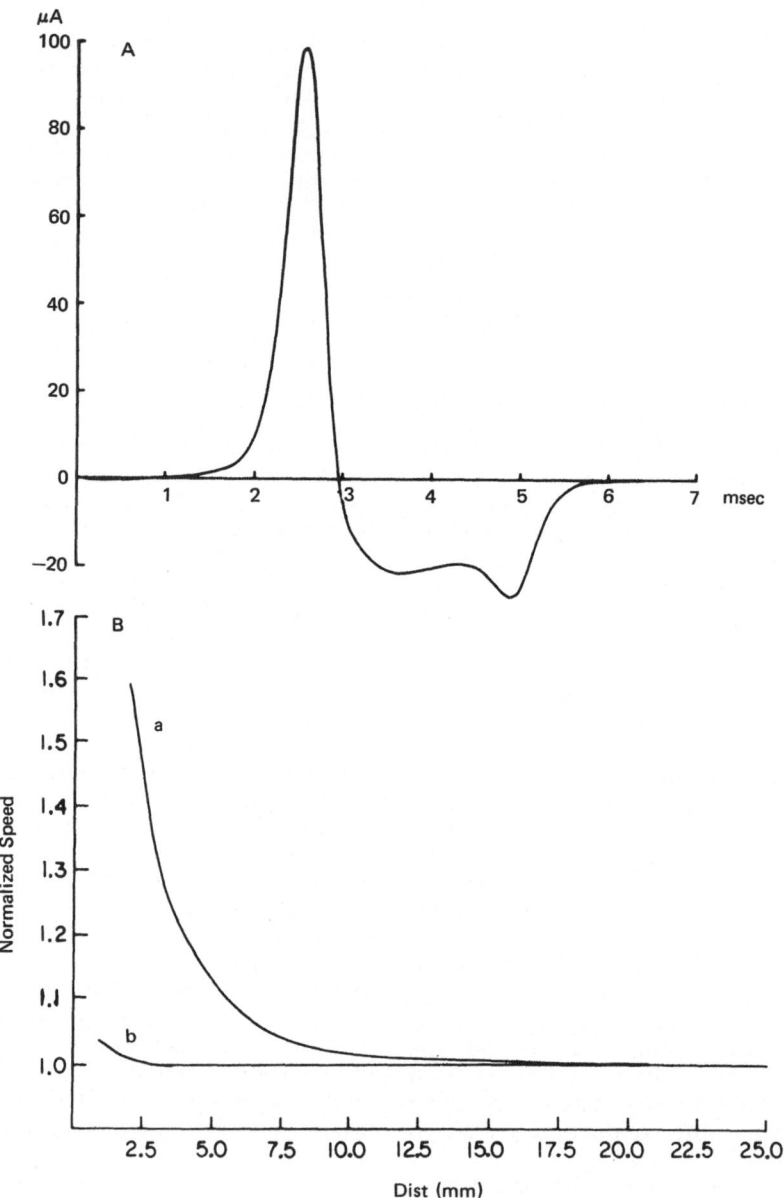

Figure 6. *A*) A plot of the current flowing axially along the cable at any given segment. A sum of exponentials fit to this shape was used as an internally applied stimulus to obtain the plot of Fig. 6*B*. *B*) Plots of normalized speed of propagation for a cable stimulated with a square pulse of current (*a*) and with the sum of exponentials fit to the axial current shown in part A (*b*). The slight difference from the constant speed of propagation at the beginning of *curve b*, was due to the failure of a perfect fit to the last portions of the axial current.

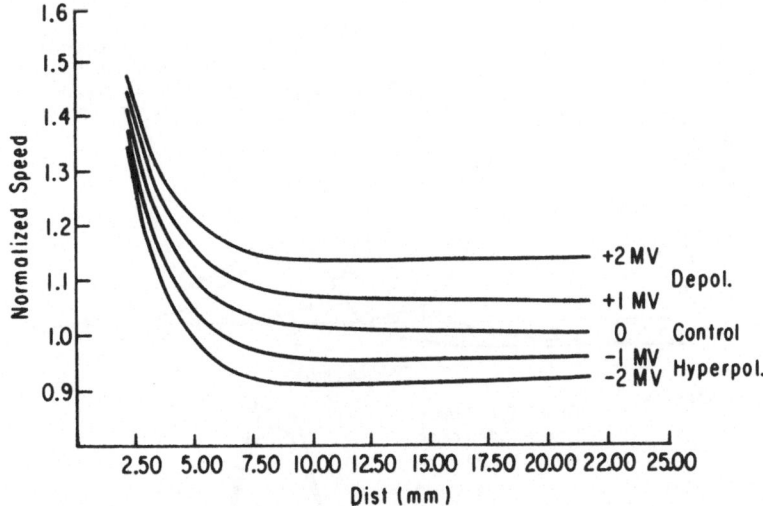

Figure 7. Plot of the normalized speed of propagation for an action potential elicited in a cable at rest, or with 1 and 2 mV hyper- or depolarization. Note that slight depolarizations increase the speed of propagation of the action potential.

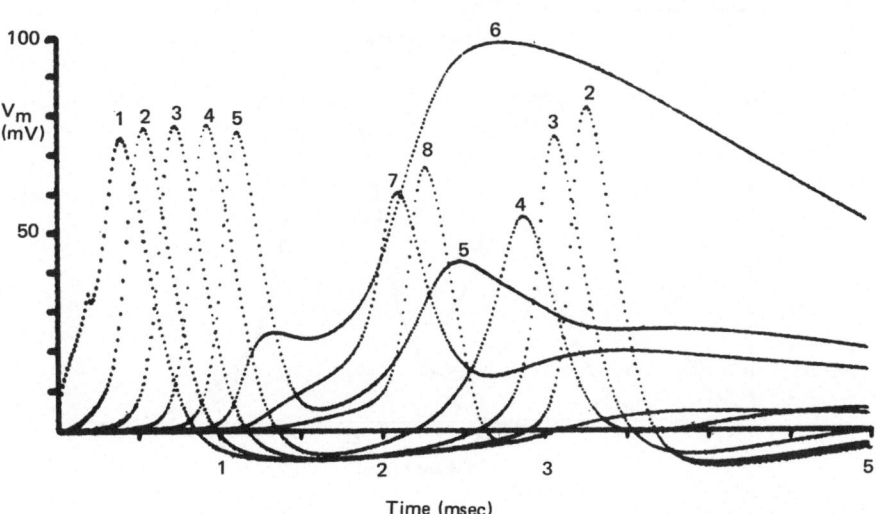

Figure 8. Plots of transmembrane potential versus time for a cable (25 C) with a region of low (3 C) temperature. The action potential travels in the warm region (*curves 1, 2, 3, 4* and *5*), reaches the cold region (*curve 6*), jumps across it (*curves 7* and *8*) and also propagates in the reverse direction (*curves 5, 4, 3* and *2* on the right of the figure). Two action potentials were obtained traveling in the reverse direction; the second one is not shown in the figure.

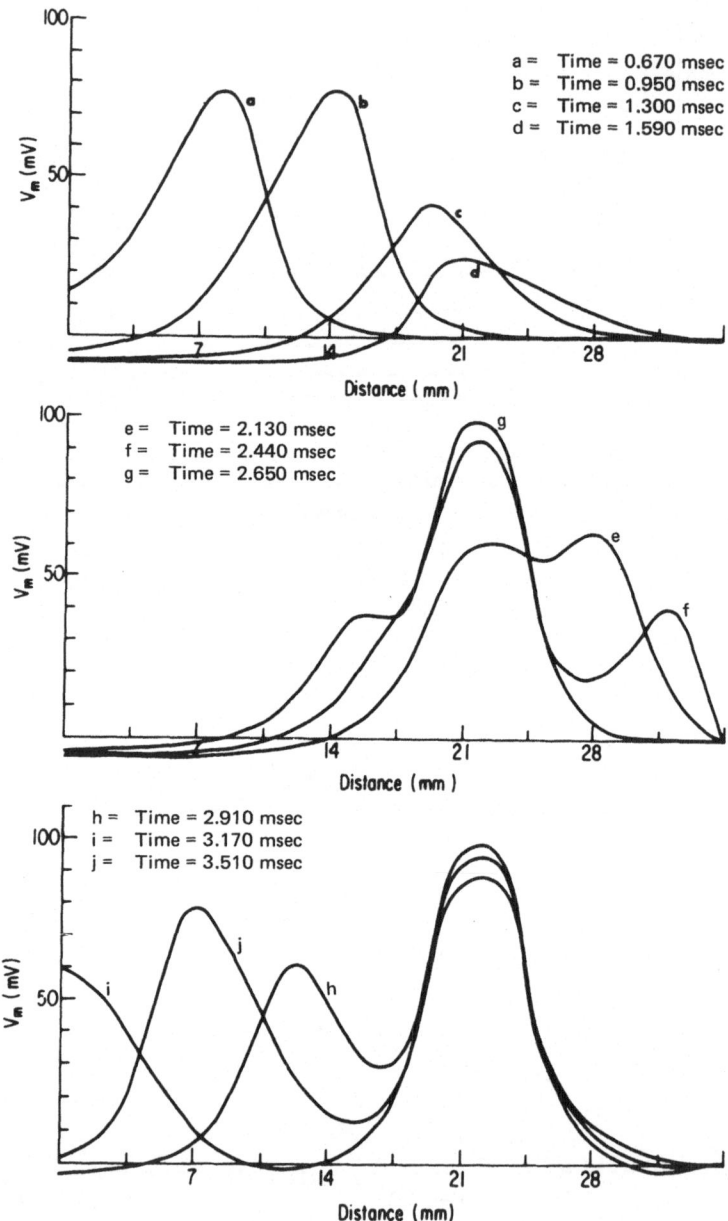

Figure 9. Plot of transmembrane voltage versus the cable distance for a cable under the same conditions as for Fig. 8. The action potential reaches the cold region at 1.590 msec after the stimulation ends, and at 2.910 msec is seen traveling in the reverse direction. The action potential at the right hand side of the figure is smaller due to the effect of the cable termination.

1.3 msec (*curve c*). The delay in the cold region allows the warm region to the right to develop an action potential (*curve d*, 1.590 msec). While the action potential in the cold region is still developing, in the warm region ahead of it a spike has already taken off and starts traveling toward the right of the figure.

Diameter differences—experiments

An example of inhomogeneities in an axon is a region of increased diameter. This condition can be reproduced in the squid giant axon in the following way. A procedure frequently used to voltage clamp the squid giant axon is to introduce an internal wire. By proper treatment, the surface resistance of the wire can be decreased to a very low value and it can effectively short-circuit a segment of the axon. Since under this condition the resistance of a smaller volume of axoplasm is in parallel to the very low resistance wire, the equivalent internal resistance of the axon has been effectively decreased. The net effect is the same as if the axon's diameter had suddenly increased, the amount dependent on the wire surface resistance.

Following the insertion of an axial wire to simulate a cell soma joining an axon another electrode was introduced through one of the axon branches, to allow us to record the internal voltage. This electrode is completely isolated except at its tip, and moving it backwards or forwards, the potential at any region inside the axon can be recorded. Stimulation of the other axon branch (Fig. 1) was accomplished through a suction electrode.

Figure 10 shows the records of the action potential at the specified points along the wire region. On the Figure it can be seen that when the recording electrode is at distances smaller than 300 µm from the axial wire tip, the action potential shows modifications. There is broadening of the action potential as it approaches the axial wire tip region as well as a decrease in velocity. In the region around the tip of the wire, the action potential becomes smaller, then shows two peaks and recovers its normal shape some distance after it enters the axial wire region.

The presence of two peaks of the action potential entering the axial wire region can be explained by the effective low internal resistance of the axon. A low internal resistance causes a much larger surface membrane to be tightly coupled and to be depolarized almost simultaneously. In this case total depolarizing current from the normal region must be distributed over a much larger capacitance, resulting in a much smaller discharge rate. This reduces the rate of rise and height of the action potential in the region adjacent to the axial wire. If the potential in the membrane surrounding the wire reaches threshold an action potential is generated there and this in turn again discharges the membrane capacitance in the region adjacent to the axial wire, giving rise to the second "hump" whose size depends on the degree to which the previous action potential has inactivated the sodium conductance.

Diameter differences—simulations

This example of axon inhomogeneity can be represented by the simulations for either an actual increased diameter or for an axial wire to represent a region of increased diameter. For the simulations shown here, the cable diameter was increased abruptly; therefore, in the numerical solution we had to deal only with the boundary condition between two homogeneous cables. A slightly different cable equation would have to be used if the diameter were changed progressively (1, 14).

When a cable has a region of increased diameter, the behavior of this region affects the propagated action potential in a complicated manner. The action potential is elicited by a current pulse, and after a slight delay, travels at a constant speed toward the region of increased diameter, whose effects on the action potential are dependent upon the change in diameter.

Figure 11 shows the effect of an increase in diameter for final ratios (in each case the later half has the same diameter) of 1 (Fig. 11*A*), 4 (Fig. 11*B*), 6.25 (Fig. 11*C*) and 10 (Fig. 11*D*). In general, the effect is that the action potential slows down as it approaches the region of increased diameter and, at the boundary, it has pronounced changes in shape that depend upon the ratio of the diameters of the two regions. For large ratios (e.g., more than 5) the action potential shows two distinct peaks. Once the action potential enters the region of increased diameter its speed increases and the original shape is recovered. Ratios of diam-

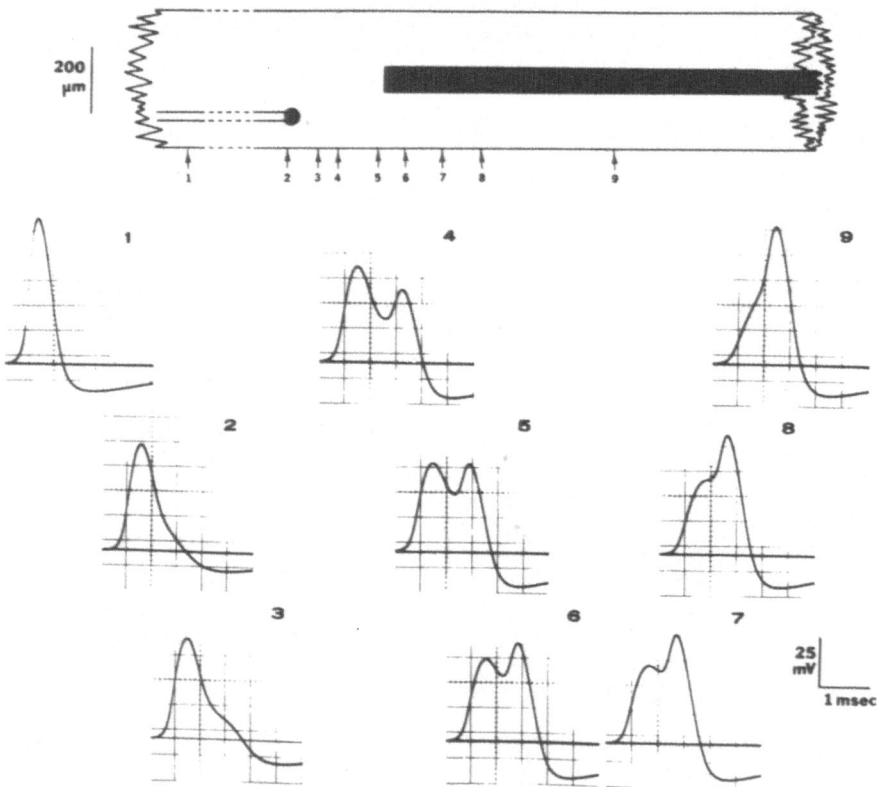

Figure 10. Experimental records obtained from a squid axon under the conditions described in the text and diagrammed in Fig. 1. The small arrows point to the axon region from where the records were obtained. The calibration in μm applies also to the distances between small arrows. The record labeled 1 was obtained about 15 mm from the axial wire.

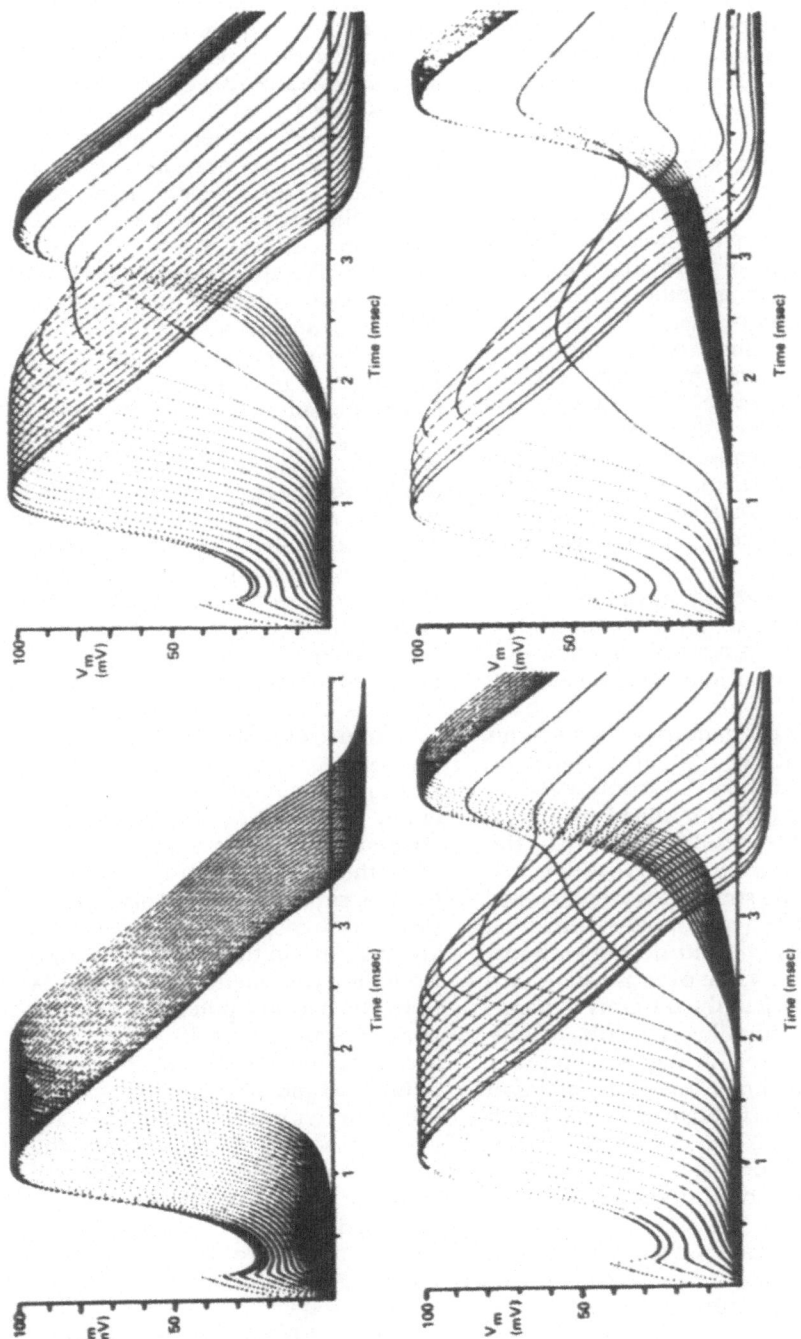

Figure 11. Plots of transmembrane voltage versus time for cables with a region of increased diameter. The ratios of the diameter of the normal cable to that of the region of increased diameter are: 1 (*A*), 4 (*B*), 6.25 (*C*) and 10 (*D*). Each successive curve corresponds to a distance of 0.5 mm on the cable. Note the decrease in the speed of propagation of the action potential as the action potential approaches the increased diameter region, and the action potential with two peaks at the boundary.

eters between the two regions greater than 3 produced also a rebound action potential that now travels in a reverse direction, toward the point of stimulation (not shown).

DISCUSSION

The experiments and simulations reported in this paper were conducted with the aim of studying the effect of axon inhomogeneities on the propagated action potential. In this paper we have described some of the effects that an excitable membrane region, affected by some "alteration," may have on another "normal" region. As examples we chose some of those that represent particular phenomena described for excitable tissues and that can be analyzed experimentally on giant axons.

Before studying propagation of action potentials on inhomogeneous axons, it is necessary to know the propagation in homogeneous axons. Fortunately much is known about normal propagation; however, an aspect that we thought would be important, is how the action potential reaches its constant speed of propagation. Our results show that the shape of the velocity curve is dependent upon the stimulus waveform. For rectangular current pulses, the initial velocity is high and declines toward the constant value over a distance up to 5 resting length constants depending on the stimulus strength. A similar result, for a digital computer simulation of myelinated nerve action potentials, was obtained by FitzHugh (8) who also found that the initial speed of the propagated action potentials is dependent upon the stimulus strength.

The axon inhomogeneities used in this paper were a) a region of different temperature and b) a region of increased diameter. The effect of a "cold" axon region on a normal region, as shown in Figs. 8 and 9, has relation to some theories proposed to account for the repetitive firing seen in some excitable tissues. An example of this effect is in the so-called "re-entry theory" of heart muscle (4), where it has been postulated that, in a normal tissue, regions may exist where the electrical behavior is affected in a similar way as that one shown in the simulations. These regions have altered their normal kinetics in such a way as to produce a very slow propagation speed or very prolonged regenerative changes. The action potential entering in one of these regions will remain there for such a prolonged period of time as to allow for the recovery of the normal regions, stimulating them again. A new stimulation of the normal regions would produce a new wave of excitation. In the examples shown in this paper, only two new excitations were elicited, but it is conceivable that short trains could be produced. Besides, there is also the possibility of the coexistence of more than one of these regions.

Another example chosen for simulation was that of a region of increased diameter, such as the change from the axon to soma in nerve cells (25). A similar situation also can be produced experimentally by inserting an axial wire in the squid giant axon. Both the experiments and the simulations show action potentials with two peaks. Similar results have been shown in the simulation of a motor neuron (7) and in the recordings from the giant neuron of *Aplysia* (25). The data presented here suggest that these action potentials with two peaks, along with similar action potentials recorded in other excitable tissues, can be accounted for by the same mechanism; namely, changes in the axial resistance.

In regard to our original objective, the study of the modifications that an action potential suffers along its way in an excitable tissue, we can conclude the following:

1) Excitable regions, affected in such a way as to produce a sustained depolarization, can give rise to new fronts of excitation that may propagate.

2) Action potentials, propagating in structures that can be treated as cables, suffer pronounced changes in shape and velocity when they reach regions of different internal resistance. One of these regions could be the area that joins an axon to the soma cell.

3) Experiments that reproduce important features of the electrical behavior of other excitable tissues can be performed on squid giant axons.

We would like to acknowledge contributions of Dr. Nelson Arispe in the early stages of this project. We appreciate the contributions of Mr. E. M. Harris (in maintaining the equipment operational) and Mrs. D. Munday (in typing the several drafts of this paper).

REFERENCES

1. ADRIAN, R. H., W. K. CHANDLER AND A. L. HODGKIN. The kinetics of mechanical activation in frog muscle. *J. Physiol. London* 204: 207–230, 1969.

2. ADRIAN, R. H., W. K. CHANDLER AND A. L. HODGKIN. Voltage clamp experiments in striated muscle fibers. *J. Physiol. London* 208: 607–644, 1970.

3. COOLEY, J. W., AND F. A. DODGE. Digital computer solutions for excitation and propagation of the nerve impulse. *Biophys. J.* 6: 583–599, 1966.

4. CRANEFIELD, P. F., H. O. KLEIN AND B. F. HOFFMAN. Conduction of the cardiac impulse. I. Delay, block, and one-way block in depressed Purkinje fibers. *Circulation Res.* 28: 199–219, 1971.

5. CRANK, J., AND P. NICOLSON. A practical method for numerical evaluation of solutions of partial differential equations of the heat-conduction type. *Proc. Cambridge Phil. Soc.* 43: 50–67, 1947.

6. DODGE, F. A., AND J. W. COOLEY. Excitation and propagation of impulses in various non-uniform axons. *Biophys. Soc. Abstr.* (Abs TPM-F5) 15th Annual Meeting, New Orleans, LA, 1971.

7. DODGE, F. A., AND J. W. COOLEY. Action potential of the motor neuron. *IBM J. Res. Develop.* 17: 219–229, 1973.

8. FITZHUGH, R. Computations of impulse initiation and saltatory conduction in a myelinated nerve fiber. *Biophys. J.* 2: 11–21, 1962.

9. HODGKIN, A. L. The subthreshold potentials in a crustacean nerve fibre. *Proc. Roy. Soc. London Ser. B* 126: 87–121, 1938.

10. HODGKIN, A. L., AND A. F. HUXLEY. A quantitative description of membrane current and its application to conduction and excitation in nerve. *J. Physiol. London* 117: 500–544, 1952.

11. HODGKIN, A. L., AND W. A. H. RUSHTON. The electrical constants of a crustacean nerve fibre. *Proc. Roy. Soc. London Ser. B* 133: 444–479, 1946.

12. KHODOROV, B. I., AND YE. N. TIMIN. Theoretical analysis of the mechanism of conduction of a nerve pulse over an inhomogeneous axon. III. Transformation of rhythms in the cooled part of the fibre. *Biophysics* 15: 503–511, 1970.

13. KHODOROV, B. I., YE. N. TIMIN, S. YA. VILENKIN AND F. B. GUL'KO. Theoretical analysis of the mechanisms of conduction of a nerve pulse over an inhomogeneous axon. I. Conduction through a portion with increased diameter. *Biophysics* 14: 304–315, 1969.

14. LIEBERMAN, M., J. M. KOOTSEY, E. A. JOHNSON AND T. SAWANOBORI. Slow conduction in cardiac muscle. A biophysical model. *Biophys. J.* 13: 37–55, 1973.

15. LORENTE DE NÓ, R. A study of nerve physiology. In: *Studies from The Rockefeller Institute, N.Y.* Vols. 131, 132, 1947. Rockefeller Institute, New York.

16. MARKIN, V. S., AND V. F. PASTUSHENKO. Spread of excitation in a model of an inhomogeneous nerve fibre. I. Slight change in dimensions of fibre. *Biophysics* 14: 316–323, 1969.

17. MOORE, J. W., AND K. S. COLE. Voltage clamp techniques. In: *Physical Techniques in Biological Research*, edited by W. L. Nastuk. New York: Academic, 1963.

18. MOORE, J. W., F. RAMÓN AND R. W. JOYNER. Axon voltage clamp simulations: I. Methods and Tests. *Biophys. J.* 15: 11–24, 1975.

19. PICKARD, W. F. On the propagation of the nervous impulse down medullated

and unmedullated fibers. *J. Theoret. Biol.* 11: 30–45, 1966.

20. PLONSEY, R. *Bioelectric Phenomena.* New York: McGraw-Hill, 1969.

21. RAMÓN, F., J. VERGARA AND J. W. MOORE. Changes in speed of propagation of action potentials in squid giant axons: Experimental and computed results. Biophysical Soc. 17th Annual Meeting, Columbus, Ohio, 1973.

22. ROSENFALCK, P. *Intra- and Extracellular Potential Fields of Active Nerve and Muscle Fibers: A Physico-mathematical Analysis of Different Models.* Copenhagen: Akademisk Forlag, 1969.

23. SABAH, N. H., AND K. N. LEIBOVIC. The effect of membrane parameters on the properties of the nerve impulse. *Biophys. J.* 12: 1132–1144, 1972.

24. STEIN, R. B., AND K. G. PEARSON. Predicted amplitude and form of action potentials recorded from unmyelinated nerve fibres. *J. Theoret. Biol.* 32: 539–558, 1971.

25. TAUC, L. Site of origin and propagation of spike in the giant neuron of Aplysia. *J. Gen. Physiol.* 45: 1077–1097, 1962.

Principles and applications of fluctuation analysis: a nonmathematical introduction[1]

CHARLES F. STEVENS

Department of Physiology and Biophysics
University of Washington School of Medicine
Seattle, Washington 98195

ABSTRACT

The mechanisms underlying many of the processes studied by membrane biophysicists are inherently probabilistic, and therefore exhibit random fluctuations around the mean of behavior. These fluctuations reflect the underlying probabilistic mechanism and therefore can sometimes provide information, not otherwise available, about these mechanisms. Fluctuations may be characterized by their spectra which are obtained from a Fourier analysis of the experimental records. When a theory for membrane processes is available, it makes predictions about fluctuation spectra and therefore may be tested by examining these spectra. Theories about gating behavior at the frog neuromuscular junction have been tested in this way, and it has been possible, in addition, to estimate the conductance of one open channel, a quantity not susceptible to direct measurements. Various physical pictures are capable of yielding the same macroscopic behavior for axon membranes, that is, the Hodgkin-Huxley equations, but these various mechanisms predict that the current fluctuations around their mean values should have different characteristics. Fluctuation analysis may, then, be of value in elucidating the physical basis for axon conductance changes.—STEVENS, C. F. Principles and applications of fluctuation analysis: a nonmathematical introduction. *Federation Proc.* 34: 1364–1369, 1975.

Over the past 5 years we have been engaged in experiments to determine the molecular mechanisms responsible for permeability changes induced by acetylcholine at the frog neuromuscular junction (1, 4, 10, 11). In addition to its relevance for synaptic physiology, we view this prepara- tion as a model system in which to study molecular mechanisms for channel gating, and hope that the information gained here will ulti-

[1] This investigation was supported by Public Health Service Grants NS 05934 and NS 10492.

mately prove of use in understanding gating phenomena in the electrically excitable membranes. Although our conclusions are of course still tentative, a very specific picture of the events associated with acetylcholine-induced gating has emerged from this work. According to our theory, two acetylcholine molecules bind onto a receptor, and this binding in turn induces a conformational change in a gating macromolecule which opens an endplate channel to permit the transmembrane flux of sodium and potassium ions. The channel thus has only two conductance states, open and closed. Transition between these conductance states is governed by rate constants α (for closing open channels) and β (the rate constant for channel opening). The rate constants themselves depend exponentially on membrane potential and this exponential dependence reflects the coupling of membrane electric field to gating through a dipole moment change of the gating molecule: a dipole moment change of at least 50 D occurs as the gate makes a transition from its open to its closed conformation. The average length of time a channel remains open depends, as a consequence of these dipole moment changes associated with gating, exponentially on membrane potential, with a doubling of open lifetime for about each -100 mV change in voltage. Typically a channel stays open, with a conductance of 20 or 30 pmhos, for about 1 msec.

Although the conclusions just presented are based on a variety of different experiments from our laboratory as well as from others, an important tool in our studies has been a technique called *fluctuation analysis*. Fluctuation analysis is a means of extracting information about the system under study from the spontaneous random fluctuations of variables being measured, that is, from the noise produced by the system. Katz and Miledi (7) discovered that a constant dose of acetylcholine applied to the frog neuromuscular junction gives rise to rather substantial random fluctuations in endplate current around its constant average value, so that endplate mechanisms, as a specific example, may be investigated with this technique. Fluctuation analysis, then, would in this instance be aimed at extracting information about underlying permeability mechanisms by examining the statistical structure of these endplate current fluctuations.

Because the technique of fluctuation analysis is rather unfamiliar even to many membrane biophysicists who use other quantitative tools regularly, it seemed to me that a presentation here of the main ideas underlying this method might be useful as an introduction.

The following discussion begins with a description of sources for fluctuations and continues with a survey of methods—mostly Fourier analysis—for characterizing these fluctuations. Finally, the techniques for interpreting fluctuations will be discussed. Throughout, I will attempt to define the vocabulary used by workers in this field so that the reader not already familiar with the terminology can approach research articles using fluctuation analysis. The discussion will make two main points: first, fluctuation analysis is a useful supplement to other methods, and can in some instances provide information that is in practice not now obtainable in any other way. Second, extracting information from fluctuations depends entirely on a theoretical framework within which the fluctuation phenomena are interpreted, and any conclusions derived from this type of investigation depend upon the validity of the theory; unfortunately, no automatic way for obtaining valid theoretical frameworks is available.

ORIGIN OF FLUCTUATIONS

At the molecular level, events are probabilistic. That is to say, the transition of a particular molecule from one state to another state occurs at random because of haphazard collisions between this molecule and others in its environment. A membrane-bound surface protein, for example, is subject to collisions by water molecules and the fluctuating energy delivered to this protein can cause random transitions between distinct conformations.

The random nature of molecular processes is revealed as fluctuations. As a specific example, imagine a population of membrane channels that can exist in either an open or a closed state, and suppose that the transition between the open and closed states of these channels occurs when a gating molecule undergoes a conformational change. If, under a specified set of experimental conditions, an average of half the channels were open and half closed, the exact fraction of open channels would from moment to moment fluctuate around 0.5. The probabilistic processes that determine conformational changes of the gating molecules give rise to these fluctuations; because of collisions between the gates and surrounding molecules, it is as if each channel repeatedly flipped a coin to decide whether or not to remain open.

The basic idea behind fluctuation analysis is that the same molecular processes that govern fluctuations are also responsible for macroscopic behavior.

CHARACTERIZATION OF FLUCTUATIONS

Before turning to some specific examples of fluctuation analysis, it will be necessary to provide some background information about how fluctuations are characterized.

Probability distribution of fluctuations

One way of characterizing fluctuations is in terms of their overall amplitude; the usual approach is to make a histogram of a fraction of the time a given fluctuating quantity has a particular value. For example, suppose that endplate current fluctuations around a mean value of 100 nA were to be studied. The endplate current would be measured a large number of times (say 1,000) and a histogram would be prepared showing how often the endplate current was within 1 nA limits at, for example, ±1 nA, ±2 nA, ±3 nA, etc., of the mean value. The amplitude of fluctuations around the mean is often specified by the histogram width as measured by its standard deviation. This standard deviation is calculated by squaring deviations from the average value (squaring gets rid of minus signs so that nothing can cancel), averaging the squared deviations, and finally extracting the square root.

It might be thought that a single number such as the standard deviation would provide only a rather impoverished description of an entire probability distribution, but this often is not the case. Many, if not most, probability distributions encountered in practice fall along the normal or Gaussian curve and this curve is entirely characterized by its mean or central value, and by its standard deviation.

In summary, then, the amplitude of fluctuations is generally specified as the standard deviation of variation around the mean value. For a normal distribution, 68% of the fluctuations fall within ±1 SD, 95%

within ±2 SD, and 99.7% within ±3 SD. If an endplate current had an average value of 100 nA and a standard deviation of 2 nA, one would observe values larger than 102 nA about 34% of the time, but an endplate current as large as 110 nA would almost never be seen.

Temporal characterization of fluctuations

If a process varies in time, clearly a characterization of its properties requires not only information about the amplitude of deviations from the mean, but also the rapidity with which such deviations occur. The rapidity with which deviations around mean value occur is usually specified either by the *covariance function* or the *spectrum*. Because the spectrum is, with modern computers, more easily and rapidly calculated, it is now the most commonly used characterization.

Fourier analysis

To understand how the spectrum of a fluctuating variable is calculated, it is necessary to review Fourier analysis. According to Fourier's theorem, functions (of the type most commonly encountered in practice) can be decomposed into a sum of sine and cosine waves of various frequencies; that is, if a 1 sec record of endplate current were subjected to Fourier analysis, the result would be weights assigned to sine and cosine functions of frequencies 1/sec, 2/sec, 3/sec, etc., so that when the sines and cosines were multiplied by these weights and added together the original record would be reconstructed. Since the same information is contained both in the original record and in the weights assigned to the sine and cosine waves, the weights constitute a complete characterization of the original record.

The conversion of a function of time into weights for sine and cosine waves is called a *Fourier transform*, and the resynthesis of the original function from the weights is termed an *inverse Fourier transform*; both Fourier transforms—that is, the characterization of a time function in terms of weights for sinusoids—and the inverse Fourier transform are very important in mathematical analysis generally and for probability theory in particular.

It was stated above that a function of time may be decomposed into a sum of weighted sines and cosines, but the precise number of sine and cosine frequencies required for this decomposition was not specified. In theory, an infinite sum is generally required; that is, one must include weights for sine and cosine waves of all conceivable frequencies. In practice, however, only a relatively small number of sine and cosine frequencies is required because the sample to be Fourier transformed is not infinitely long and because the recording equipment has a finite bandwidth. The lowest frequency (other than the average DC level) used in a decomposition is set by the sample length. Thus, the lowest frequency component for a 1 sec sample would be a 1 Hz sine (and cosine) wave and for a ½ sec sample the lowest component would be a 2 Hz sine (and cosine) wave. The highest frequency is, in practice, generally provided by the rate at which the original time record was sampled into the computer that is used to perform the Fourier analysis: the highest frequency component is, according to the *sampling theorem*, ½ of the sampling rate. Thus, if a waveform were sampled at 1 kHz, the highest frequency component in the Fourier analysis would be 500 Hz, and if the sampling frequency were 4/sec, the highest frequency component would be a sinusoid with the frequency 2 Hz. Weights for sines and cosines are thus multiples of one period per record up to the highest component determined by the sampling rate (via the sampling theorem). For example, if a 1 sec record were taken at a rate of one sample point per msec, Fourier analysis would yield weights for sine waves (and cosine waves) of 1/sec, 2/sec, 3/sec, and so forth, up to 500/sec. If a ½ sec sample were taken at a sampling rate of 4 kHz, the sinusoid components would have frequencies of 2/sec, 4/sec, 6/sec, and so forth up to 2,000/sec.

Calculation of spectra

The spectrum of an experimental record is calculated in the following way: the record is Fourier analyzed to convert the original time function into the equivalent characterization in terms of weights for sine and cosine function. For each frequency component, the weights of the associated sine and cosine functions are squared and averaged (that is, squared, added together, and divided by 2). This collection of average squared weights for the various frequency components constitutes the spectrum for the original process. For example, if a 1 sec long sample of fluctuating endplate current were sampled at a rate of 1 kHz, this record would be Fourier analyzed and squared weights associated with the 1, 2, 3, 4 Hz, etc. up to 500 Hz sinusoids would constitute the spectrum for the original process.

The spectrum associated with a record characterizes the rapidity of fluctuations in the original record. If the measured variable fluctuated at a very high rate, its Fourier decomposition would contain very many high frequency components so that its spectrum would have relatively large values for the high frequencies. If, on the other hand, the original fluctuation were slowly varying, the higher frequency component weights would be small and only the slow component weights would be appreciable. The spectrum of a sine or cosine wave has only a single point at the sinusoid's frequency. At the other extreme is white noise whose spectrum has equal values at all frequencies. In some cases fluctuations are named according to their spectral characterizations: $1/f$ noise has spectral amplitudes that are inversely proportional to frequency so that, for example, the 1 Hz component is 10 times larger than the 10 Hz amplitude.

Information contained in spectra is often most easily grasped when presented in a graphical form: generally, log spectral amplitude is plotted against log frequency so that both the amplitude and the frequency axes are compressed to contain information over many orders of magnitude.

Spectra may be calculated in a variety of ways. One simple way is to tape-record the original signal and play the tape over many times, first through a band-pass filter with a center frequency of 1 cycle, second with 2 cycles, third with 3 Hz and so forth. The output of the band-pass filter is measured with an rms meter and the square of the rms meter reading gives the spectral amplitude for the filter center frequency. In practice this procedure is too cumbersome, and although once standard, has now been replaced by techniques that employ the digital computer. The technique almost always employed now for Fourier decomposition is the *Fast Fourier Transform*, usually abbreviated FFT, an algorithm that decreases the amount of time required for calculations by making clever use of inherent properties of digital computers. On a small lab computer, the spectrum of a 1,000 point sample of a fluctuating variable may be calculated in approximately 1 sec.

Of the many errors that may be made in recording and interpreting spectra, two deserve special mention. First, the spectrum of a fluctuating process must be calculated from a relatively large number of samples to provide a reliable characterization of the fluctuations. Because processes that generate fluctuations are inherently random, the amplitudes of spectral estimates can vary quite considerably from one record to the next. Thus, in order to obtain reliable estimates, one usually averages the spectra obtained from a number of different records. The accuracy of the spectral estimates depends, for a given sample length and sample frequency, on the number of spectra averaged together, so that doubling the number of spectra in an average generally doubles the precision of the spectral estimates.

Second, care must be taken to avoid a type of error known as *aliasing*. To appreciate how aliasing arises, consider a situation in which a sample rate of 1,000 cycles/sec is being used. If a 500 Hz sine wave (the upper frequency limit possible, according to the sampling theorem) is sampled at this rate, precisely two samples per cycle will be obtained; suppose for convenience that these two samples occur at the peak and trough of a sine wave. Now, if a sine wave of 1,500 Hz, that is, three times the frequency, is sampled in the same way, the sample point will fall again on peaks and troughs of the sinusoid to yield the sparsely sampled record which looks identical to the one obtained from the 500 Hz sine wave. Thus, in general, if sine waves of higher frequencies than one-half the sample rate are present in the record

being prepared for processing, these higher frequency components can give rise to sample points that are just what one would get from a low frequency sine wave. If the sampled record is now Fourier analyzed with, for example, the Fast Fourier Transform, the high frequency components —those above the range that can be reconstructed by the Fourier analysis technique—will be treated by the Fourier analysis program as if they were low frequency components, thereby giving rise to spuriously high values for some of the lower frequency spectral amplitudes. This phenomenon of higher frequency sinusoids being confused with lower frequency sinusoids is known as aliasing and can be an extremely important source of error. Aliasing may be eliminated by filtering out the higher frequency components with an analog filter before the record is sampled. Thus, if a record is to be sampled at 1 kHz, the highest frequency that will appear in the Fourier analysis is 500 cycles and any frequency components in the original record above 500 cycles must be filtered out to prevent aliasing.

Calculation of the covariance

The second method of characterizing the temporal properties of a fluctuating signal is through the *covariance* or *autocorrelation function*. The covariance function may be calculated directly from the record under investigation, but generally it is more efficient, if a computer is available, to calculate the covariance from the spectrum by a Fourier transform: the covariance function is the inverse Fourier transform of the spectrum. Thus to calculate a covariance function for a specific record, one would first obtain the spectrum by the FFT method. Then, the spectrum would be Fourier transformed by adding together cosine waves[2] of various frequencies weighted according to the corresponding spectral amplitudes. If the spectrum had estimates at 1, 2, 3, up to 500 Hz, cosine waves of 1, 2, 3, up to 500 Hz frequencies would be weighted by the corresponding spectral amplitudes and added together. The resulting function of time would be the required covariance function. If this covariance function is normalized so that its value at time = 0 is 1, it is known as the autocorrelation function.

Just as the spectrum contains information about the rapidity of fluctuations in the original record, so does the covariance function provide, in the time domain, a similar picture. Specifically, the covariance function $C(t)$ indicates how well the current value of a signal will be correlated with its value t seconds later. Usually covariance functions for random processes decline more or less smoothly from a maximum value at time = 0 to a value of 0 for long times. If this decay occurs over a 1 msec period, this would mean that the original fluctuating signal had correlations lasting about 1 msec, whereas a covariance function that declines over 100 msec would indicate that the original signal was more slowly varying and was therefore significantly correlated at times up to about 100 msec. Covariance functions often give a more easily grasped picture of the temporal structure of a random process because they illustrate in the time domain the rapidity of fluctuations. Although it is not obvious from the preceding description, the 0 time value of a covariance function is nothing more than the square of the

standard deviation (described earlier) of the original record.

In summary, fluctuations—that is, fluctuating functions of time—are characterized by the magnitudes of the fluctuations, usually given as the variance or standard deviation (rms value), and the rapidity of fluctuations as described by the covariance function or the spectrum. We now turn to the use of these characterizations for yielding information about the system under study.

INTERPRETATION OF FLUCTUATIONS

Useful information can be obtained from fluctuations only through some connection between the properties of the fluctuation and macroscopic behavior of the system studied. Two approaches to establishing this connection, the fluctuation-dissipation theorem and explicit mechanistic theories, will be considered now.

Fluctuation-dissipation theorem approach

According to the fluctuation-dissipation theorem, the covariance function associated with a system's fluctuations around a constant mean describes relaxations of the system when it is perturbed. A use of the fluctuation-dissipation theorem is illustrated by the following specific example: Suppose that a technique could be devised whereby the concentration of acetylcholine at a neuromuscular junction could be changed stepwise from one level to another. A stepwise increase in acetylcholine concentration would cause additional endplate channels to open, and this relaxation to the new value of endplate conductance would, according to our observations, follow an exponential time course. Thus, the relaxation of number of channels open from one value to another—

---

[2] Sine waves do not occur because the spectrum is an *even* function of frequency.

caused by changes in acetylcholine concentration—would be an exponential whose rate constant we call α.

The fluctuation-dissipation theorem permits us to determine this relaxation and the value of the rate constant α without having to carry out the technically difficult operation of causing a step change in acetylcholine concentration. To use the fluctuation-dissipation theorem approach, a constant dose of acetylcholine is applied and the fluctuations of endplate conductance as a function of time around the mean value are measured. When the covariance function of these fluctuations is calculated (by determining the spectrum with the FFT and then Fourier transforming the spectrum to yield the covariance function), it is found to be of the form $e^{-\alpha t}$. According to the fluctuation-dissipation theorem, this same function describes the relaxation of endplate conductance in response to step changes in acetylcholine concentration, and the constant α in the covariance function is the rate constant for these relaxations. Thus, determining the covariance function for acetylcholine-induced endplate conductance fluctuations and applying the fluctuation-dissipation theorem is equivalent to carrying out the technically much more difficult concentration-jump experiment to measure the time course of endplate conductance response to changes in acetylcholine concentration.

Before considering applicability and limitations of the fluctuation-dissipation theorem, it may be helpful to place this theorem in its context. Although the theorem, like other results in science, has a long history, the first explicit statement of the essential notion appears to have been made in papers by Callan and co-workers (2, 3); these ideas were later systematically developed and applied by Kubo (8, 9). The fluctuation-dissipation theorem forms the basis for one of the several standard approaches that permit statistical mechanical calculations to be made about the behavior of systems away from equilibrium. The problem is that, although quite detailed calculations about the equilibrium state of a system generally may be made, difficulties arise when one attempts to treat the response of the same system to perturbations away from equilibrium. By using the fluctuation-dissipation theorem, however, the response of a system to perturbations may be predicted from a knowledge of fluctuations that occur in the equilibrium. Equilibrium statistical mechanics permits calculation of the covariance function for spontaneous fluctuations around equilibrium, and this function, through the fluctuation-dissipation theorem, then yields the response for external driving; in this way, driven responses may be found from a knowledge of behavior at equilibrium.

In statistical physics, the fluctuation-dissipation theorem is of importance as one of the principal keys for studying nonequilibrium behavior of systems. In biology, where purely theoretical studies are seldom fruitful, the fluctuation-dissipation theorem's main use probably is a link between experimentally measured fluctuations and the response of a system to external forcing. Thus, when it is inconvenient or impossible to perform the experimental manipulations required for determining a system's response to external perturbations, a study of fluctuations can give the same information. Fluctuation analysis then can serve to yield, for example, the rapid response of chemical systems when standard techniques, such as temperature and pressure jumps, are not for some reason feasible. Some of the advantages of this approach have recently been discussed by Feher and Weissman (5).

In spite of the fact that the fluctuation-dissipation theorem has been proved to hold for certain situations in statistical mechanics, it is not always applicable to arbitrarily chosen systems. In fact, it is very easy to construct examples of realistic biological and chemical systems to which this theorem cannot be legitimately applied and is, in fact, incorrect.

Usually, the theorem can only be applied to a system whose future evolution depends exclusively on its present state and not on the manner in which that state was reached. For example, suppose a chemical reaction can be driven to final products by adding a particular reactant, and suppose that a variety of intermediate steps occur between the initial reaction with the added reactant and final products. If concentration of the final products is viewed as the response, and if reactant concentration is identified as the driving function, future changes in product concentration (the response) could not be predicted solely from present product concentration and reactant concentration, because the future evolution of the response would depend on the concentrations of various intermediates; the concentrations of intermediates would in turn vary with the history of product and reactant concentrations. For this system, then, one could not predict the response of products to increments of reactant concentration by studying the fluctuations around equilibrium concentration of products determined by a fixed reactant concentration; that is, this is an example of a system to which the fluctuation-dissipation may not be applied. It should be noted that if the state of this chemical system is specified by the concentrations of reactants, products, and all intermediates (and not just reactants and products), then the fluctuation-dissipation theorem could be applied to this more complicated multidimensional descriptive variable.

The applicability of the fluctuation-dissipation theorem is limited by more than the presence of hidden states. The original derivation of this theorem required that deviations from equilibrium be small, but the meaning of "small" in the biological context—and indeed in the physical context—is often vague. Essentially, the system must be perturbed within its linear range, but often no a priori knowledge about what constitutes a range of linear behavior for a complex system is available.

The fluctuation-dissipation theorem, although important and powerful, must thus be applied only with care and consideration for possible limitations, for no experimental test for applicability of this theorem is available. It should be emphasized that the fluctuation-dissipation theorem approach is, in contradistinction to the method described next, atheoretical in the sense that no specific assumption must be made about the nature of underlying mechanisms. This atheoretical nature of the fluctuation-dissipation theorem increases the generality of the method, but simultaneously limits the information about specific mechanisms that may be gained from analysis of fluctuations through the application of this theorem.

Approach via mechanistic theories

The second way to connect data from fluctuations with properties of a system under investigation is to construct a probabilistic theory for the system's operation. Probability theory then provides a standard means for calculating both the average response —that is, the microscopic behavior —as well as the characteristics of the fluctuations to be expected.

Two points deserve a special emphasis in connection with this theoretical approach. First, there is no automatic way of developing the theories that relate properties of fluctuations to other behavior of the system. The methods of probability theory provide the framework for doing so, but the specific assumptions that go into the theory must still, as usual, be arrived at through insight into fundamental mechanisms. Thus, connecting fluctuations with microscopic behavior through a theoretical approach is no different from understanding how the system works. Second, any inference made about a system's properties from analysis of fluctuations is no more accurate than the probabilistic theory used to describe the system's behavior. For example, if a chemical kineticist were to obtain a rate constant by using fluctuation analysis rather than by the more conventional temperature jump technique, this rate constant would be reliable only in so far as the probabilistic theory that described the chemical system's behavior was accurate. Thus, interpretations of spectra by a theory depend entirely on correct representation of underlying mechanisms.

A specific example may be helpful in clarifying this theoretical approach. On the basis of investigations of endplate currents evoked by nerve stimulation, it was inferred that the acetylcholine concentration increase is extremely brief and that the rate limiting step in the decay of normal evoked endplate currents is the closing of endplate channels, presumably a conformational change. These investigations lead to the following specific theory about endplate permeability increases: Acetylcholine binds (according to the law of mass action) to its receptor, and the associated channel then opens with a constant probability per unit time. Once open, the channel closes with a constant probability per unit time, α. This picture, together with the assumption that neighboring channels interact only slightly if at all, is adequate to account quantitatively for the time course of nerve evoked endplate currents. The difficulty in performing a more definitive test of this theory was, however, that the cleft acetylcholine concentration change had to be inferred from endplate currents; this concentration could neither be directly measured nor experimentally controlled because any means of applying acetylcholine other than by nerve release causes concentration changes that are too slow. (Temperature jump changes are theoretically possible but practically extremely difficult.) An alternative way to test this view, then, was to predict the fluctuations in endplate conductance from the theory whose parameters had been independently determined from measurements on evoked endplate currents, and then compare the predicted and observed spectrum for fluctuations. In this way, the theory could be used to make predictions for a case in which the cleft acetylcholine concentration was, if of unknown magnitude, at least constant. We found that the theory described above does indeed predict the endplate conductance fluctuations very accurately (Anderson and Stevens (1)).

In this instance, then, fluctuation analysis was used to circumvent the difficulty that no methods are available for rapidly controlling cleft acetylcholine concentration. Once the theoretical framework used to interpret spectra could be tentatively accepted, it also became possible to estimate the conductance of one open channel: the single channel conductance appears as a parameter in the equations derived from our theory for the spectrum of endplate conductance fluctuations, and thus may be obtained from fits of the theoretical expressions to the observed spectra. We found the single channel conductance to be 20 or 30 pmhos, although it must be emphasized that the accuracy of this value depends entirely on the validity of the assumptions used in constructing our theoretical description of endplate conductance mechanisms. Since single channel conductance is not susceptible to direct measurement, its estimation through fluctuation analysis provides another example of information that could not in practice be obtained by other means.

Most of the examples mentioned earlier have been drawn from our studies of endplate permeability mechanisms, but fluctuation analysis is not, of course, limited to this system. Another related use of these methods is in making inferences about mechanisms underlying the axonal excitability processes described by the Hodgkin-Huxley equations.

A variety of different physical models lead to the Hodgkin-Huxley equations, and these different theories can make identical predictions about the results of any standard voltage clamp experiment; two rather different physical pictures which yield the same final equations have been presented earlier (12), but construction of other examples is not difficult. Fluctuation analysis may well prove helpful in distinguishing between the various alternatives that may be proposed, however, because theories which predict identical average membrane responses to changes in voltage can (and, in general, do) make quite different predictions about the properties of axon conductance fluctuations.

According to our experimental results (Anderson and Stevens (1)),

the endplate permeability mechanisms obey the fluctuation-dissipation theorem, at least to the precision of our observations. The Hodgkin-Huxley equations are especially interesting in this respect, however, because the physical picture that might seem to provide the most plausible basis for these equations—that is, the independent no-memory gates guarding a channel with only two conductance states, as originally proposed by Hodgkin and Huxley (6) —does not conform to the fluctuation-dissipation theorem, although a variety of alternative models can be constructed that do, to various degrees of precision, conform. Because a number of laboratories are currently making measurements of axon conductance fluctuations (see articles by Fishman and DeFelice, this issue), results of these experiments should soon be able to indicate which class of theories (those that conform to versus those that depart from the fluctuation-dissipation theorem) are tenable.

REFERENCES

1. ANDERSON, C. R., AND C. F. STEVENS. *J. Physiol. London* 235: 655, 1973.
2. CALLEN, H. B., AND R. F. GREENE. *Phys. Rev.* 86: 702, 1952.
3. CALLEN, H. B., AND T. A. WELTON. *Phys. Rev.* 83: 34, 1951.
4. DIONNE, V. E., AND C. F. STEVENS. *J. Physiol. London* In press.
5. FEHÉR, G., AND M. WEISSMAN. *Proc. Natl. Acad. Sci. U.S.* 70: 870, 1973.
6. HODGKIN, A. L., AND A. F. HUXLEY. *J. Physiol. London* 117: 500, 1952.
7. KATZ, B., AND R. MILEDI. *Nature* 226: 962, 1970.
8. KUBO, R. *J. Phys. Soc. Japan* 12: 570, 1957.
9. KUBO, R. In: *The Many-Body Theory*, edited by R. Kubo. Tokyo: Tokyo Summer Institute of Theoretical Physics, 1965.
10. MAGLEBY, K. L., AND C. F. STEVENS. *J. Physiol. London* 223: 151, 1972.
11. MAGLEBY, K. L., AND C. F. STEVENS. *J. Physiol. London* 223: 173, 1972.
12. STEVENS, C. F. *Biophys. J.* 12: 1028, 1972.

Effects of cholinergic compounds on the axon-Schwann cell relationship in the squid nerve fiber

JORGE VILLEGAS

Centro de Biofísica y Bioquímica
Instituto Venezolano de Investigaciones Científicas
Caracas 101, Venezuela

ABSTRACT

The effects of acetylcholine, carbamylcholine, D-tubocurarine, eserine, and α-bungarotoxin on the Schwann cell electrical potential of resting and stimulated squid nerve fibers were studied. Acetylcholine (10^{-7} M) and carbamylcholine (10^{-6} M) induce a prolonged hyper polarization in the Schwann cells of the unstimulated nerve fiber. In the presence of carbamylcholine (10^{-6} M) the behavior of the Schwann cell membrane to changes in the external potassium concentration approximates the behavior of an ideal potassium electrode. D-Tubocurarine (10^{-9} M) blocks the hyperpolarizing effects of nerve impulse trains and carbamylcholine (10^{-6} M), whereas at the same concentration eserine prolongs the Schwann cell hyperpolarizations induced by axon stimulation or by acetylcholine (10^{-7} M). α-Bungarotoxin (10^{-9} M) also blocks the hyperpolarizing effects of nerve impulse trains and of carbamylcholine. D-Tubocurarine (10^{-5} M) protects the Schwann cells against the irreversible action of α-bungarotoxin. These results show the existence of acetylcholine receptors in the Schwann cell membrane. Preliminary measurements of the binding of ^{125}I-α-bungarotoxin to the plasma membranes isolated from squid nerves also indicate the presence of acetylcholine receptors. These findings support the involvement of cholinergic mechanisms in the axon-Schwann cell relationship previously described.—VILLEGAS, J. Effects of cholinergic compounds on the axon-Schwann cell relationship in the squid nerve fiber. *Federation Proc.* 34: 1370–1373, 1975.

Previous work carried out in the giant nerve fiber of the squid showed that whereas the propagation of a single nerve impulse has little effect on the Schwann cell electrical potential (18), the conduction of nerve impulse trains by the axon is followed by a prolonged hyperpolarization of the Schwann cells (15). Since the increase in potassium concentration in the axon-Schwann cell interspace during the nerve impulse trains (4) cannot account for these potential changes, they have been considered as the electrical expression of some sort of coupling between the axon and its satellite cells.

The presence of acetylcholine and its related enzymes in peripheral nerves is currently thought to be related to their transport towards the active physiological sites of the nerve terminals at synapses and neuromuscular junctions (7), especially after the intervention of acetylcholine in the mechanism responsible for axonal conduction was ruled out (11). However, the axon-Schwann cell interactions described above and the presence of membrane specialized regions in the axon (13), which appear at places where the trilaminar substructure of the axolemma and a narrowing of the axon-Schwann cell interspace are also observed (14), made it important to investigate the possible existence of axon-Schwann cell chemical coupling through the acetylcholine system.

The present work deals with the analysis of the effects of the cholinergic transmitter substance, curare, eserine and α-bungarotoxin on the Schwann cell and axon membrane potentials in the giant nerve fiber of the tropical squid *Sepioteuthis sepioidea* (16, 17). The unpublished work on α-bungarotoxin briefly described in this paper represents the results of experiments carried out in collaboration with R. Villegas, G. M. Villegas and F. V. Barnola, which will be published in detail elsewhere.

ACETYLCHOLINE AND CARBAMYLCHOLINE

Acetylcholine (ACh) and its analog carbamylcholine (carbachol) were used to investigate whether the long-lasting effects of nerve impulse train conduction on the Schwann cell membrane potential could be reproduced by the external application of the cholinergic transmitter substance. Acetylcholine has a higher affinity than carbamylcholine for specific receptors in the cell membrane (6). However, ACh is rapidly hydrolyzed by the acetylcholinesterase present in the tissues (10).

Figure 1 shows the effects of acetylcholine and carbamylcholine on the Schwann cell and axon electrical potentials in the unstimulated intact nerve fiber. Figure 1a shows that a brief (1 min) exposure to acetylcholine (10^{-7} M) is followed by a transient hyperpolarization of the Schwann cell. A similar effect, though more prolonged, is produced by an equally brief pulse of carbamylcholine (10^{-6} M), as may be seen in Figure 1b. In both cases, the membrane potential of the axon remained unchanged during the whole experimental period.

The nature of the ionic mechanisms underlying the hyperpolarizing effect of the cholinergic transmitter on the Schwann cells was also explored.

Figure 2 shows the relationship between the external potassium concentration and the Schwann cell membrane potential measured in axon-free nerve fiber sheaths, in the absence and in the presence of carbamylcholine (10^{-6} M) in the external medium. This figure shows that externally applied carbamylcholine increases the sensitivity of the Schwann cell membrane potential to changes in the external potassium concentration between 0.1 and 50 mM. Thus, in the carbamylcholine-treated nerve fibers the behavior of the Schwann cell membrane approximates the behavior of an ideal potassium electrode.

It was also found that either a 100-fold reduction in the external sodium concentration or the external appli-

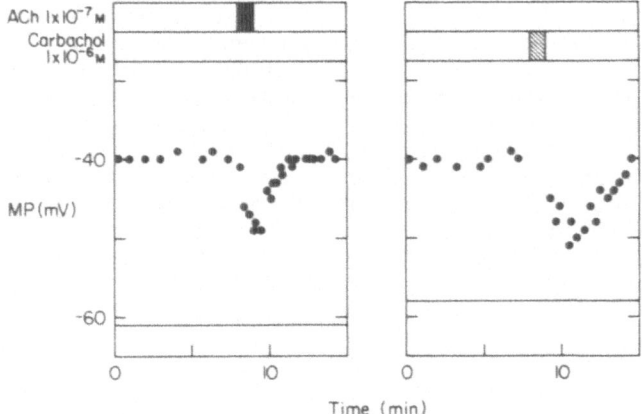

Figure 1. Effect of the cholinergic transmitter substance on the Schwann cell membrane potential of the unstimulated nerve fiber. Each graph corresponds to the results obtained in a different nerve fiber exposed to (a) acetylcholine or (b) carbamylcholine. The electrical potentials have been plotted as a function of time. Each point corresponds to the potential difference recorded in a different Schwann cell in the same nerve fiber. The axon membrane potential (line) was monitored. The striped bars indicate the intervals during which the physiological sea water solution was replaced by sea water solutions containing acetylcholine (10^{-7} M) or carbamylcholine (10^{-6} M). After a 1 min exposure to the test solution the Schwann cell became hyperpolarized by about 10 mV for several minutes and then returned gradually towards its initial potential level.

cation of tetrodotoxin (50 nM) has no appreciable effect on the Schwann cell hyperpolarizations induced by carbamylcholine (see ref 17, Figures 5 and 6). These experimental findings have been considered as indicating the presence of acetylcholine receptors in the Schwann cell membrane. These results also suggest that the external application of the cholinergic transmitter to the unstimulated nerve fiber increases the relative permeability of the Schwann cell membrane to potassium (17).

CURARE

D-Tubocurarine was used to investigate whether the long-lasting effects on the Schwann cell membrane potential of nerve impulse trains, and those of acetylcholine and carbamylcholine described above are related.

D-Tubocurarine is a competitive antagonist for the cholinergic transmitter at specific nicotinic receptor sites (5).

Figure 3 shows the effects of D-tubocurarine (10^{-9} M) on the axon and Schwann cell membrane potentials, in the intact nerve fiber before and after the stimulation of the axon (a) or the addition of carbamylcholine (10^{-6} M) to the external sea water medium (b). The graphs show that the Schwann cells become hyperpolarized within 3 min after the addition of D-tubocurarine, and that they gradually return to the initial levels within the next 5 min. The graphs also show that once the Schwann cells have recovered from their transient hyperpolarizations, the conduction of nerve impulse trains or the external application of carbamylcholine has no appreciable effect on the Schwann cell membrane potential. The resting and action potentials of the axon remained unchanged during the whole experimental periods.

It was also found that the Schwann cells in axon-free nerve fiber sheaths undergo a transient hyperpolarizing potential change after the external application of D-tubocurarine (see ref 16, Figure 6). Thus, the initial hyper-polarization induced by D-tubocura-rine in the Schwann cells of the un-stimulated intact nerve fiber repre-sents a direct action of the drug on the cell membrane.

These experimental findings have been considered as indicating that the acetylcholine receptors present in the nerve fiber are directly involved

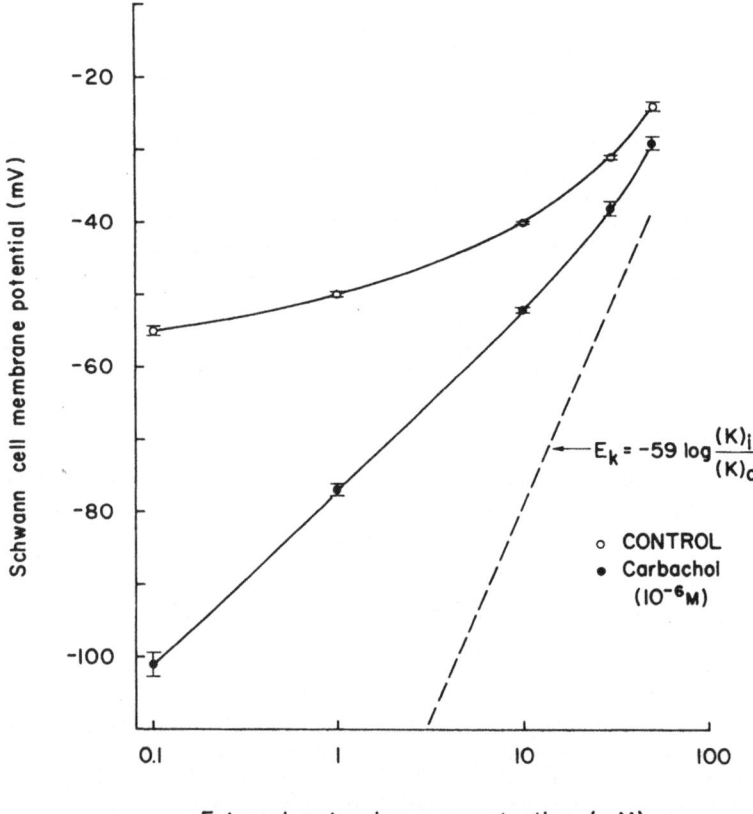

Figure 2. Effect of carbamylcholine on the relationship between Schwann cell electrical potential and external potassium concentration. Results obtained in 20 axon-free nerve fiber sheaths. The Schwann cells were impaled directly from the axonal surface of the sheaths. Each value is the mean ± SE of the potential differences measured in at least 23 (range 23–260) different Schwann cells impaled in each test solution. The dashed line indicates the behavior of an ideal potassium electrode. The open circles correspond to measurements made in control sea water solutions, and the solid circles to measurements made in sea water solutions containing carbachol (10^{-6} M). The Schwann cell electrical potential changes induced by variations of the external potassium concentration in the carbamylcholine sea water solutions are larger than those observed for the control solutions.

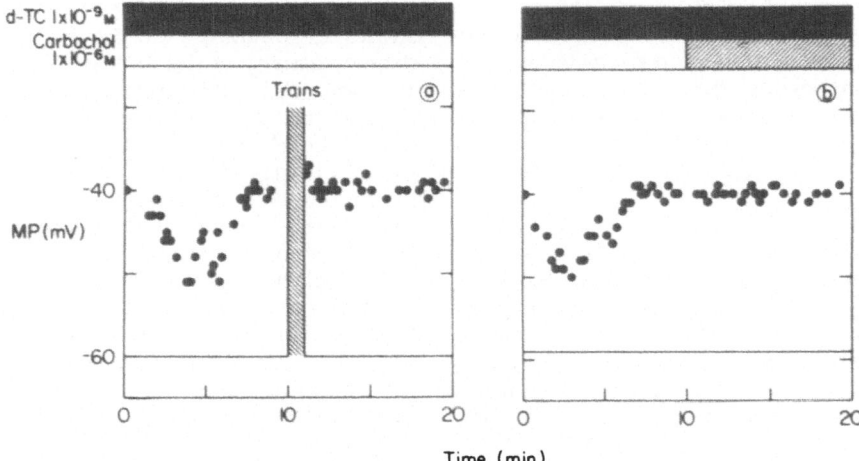

Figure 3. Effect of D-tubocurarine on the Schwann cell membrane potentials recorded before and after (*a*) prolonged stimulation of the axon or (*b*) external application of carbamylcholine. The electrical potentials have been plotted as in Fig. 1. At 0 time D-tubocurarine (shaded bar) was added to the external medium, this being followed by a transient hyperpolarization of the Schwann cell. In (*a*) stimuli were delivered to the axon at 125/sec during the interval indicated by the vertical striped bar. In (*b*) carbamylcholine was added to the bathing solution at the time indicated by the horizontal striped bar. The hyperpolarizing effects of nerve impulse trains and carbamylcholine were abolished by the incubation in 10^{-9} M D-tubocurarine.

in the mechanism responsible for the long-lasting delayed hyperpolarizations recorded in the Schwann cells around stimulated axons.

ESERINE

Eserine (physostigmine) is known to inhibit cholinesterase activity specifically (10). However, at concentrations higher than those needed to inhibit the esterases, eserine appears to also antagonize the action of acetylcholine (3).

The presence of acetylcholinesterase in *S. sepioidea* nerve fibers has been established both by histochemical electron microscopic determinations in the intact fiber (12) and by the in vitro assay of its enzymatic activity in the subcellular fractions isolated from these nerve fibers (G. Camejo and J. Villegas; J. Villegas,

Flor V. Barnola and R. Villegas, unpublished results).

Figure 4 shows the effects of eserine (10^{-9} M) on the Schwann cell membrane potential, and on the long-lasting hyperpolarizations induced in the Schwann cell either by the prolonged stimulation of the axon (*a, b*) or by the external application of acetylcholine (*c, d*). The graphs show that at this low concentration in the external medium eserine appears to prolong the hyperpolarizing effects of the nerve impulse trains and of acetylcholine on the Schwann cell electrical potential.

It was also found that even at a high concentration (10^{-4} M) in the external medium eserine had no appreciable effect on the resting and action potentials of the axon (16).

These experimental findings have been considered as indicating that the acetylcholinesterase present in these nerve fibers may be directly involved in the mechanism responsible for the long-lasting delayed hyperpolarizations observed in the Schwann cells around stimulated axons.

BUNGAROTOXIN

The purified α-toxin from the venom of *Bungarus multicintus* is known to constitute an irreversible blocking agent of neuromuscular transmission (1) and of the response of various muscle and electric organ preparations (2, 8, 9) to cholinergic agonists. Until the present it has no known effects on acetylcholinesterase,

either on the binding of the substrates or inhibitors or on the hydrolytic activity of the enzyme (2, 8). In addition, D-tubocurarine protects against the action of α-BGT (2, 8). Thus, the blocking effect of α-BGT on cholinergic systems has been attributed to its highly specific binding to acetylcholine receptor sites in the cell membrane (2, 8).

The effects of purified α-BGT (obtained as a gift from Dr. Michael Raftery, from the California Institute of Technology, USA) on the Schwann cell electrical potential of the intact or slit nerve fiber are shown in Figures 5–7.

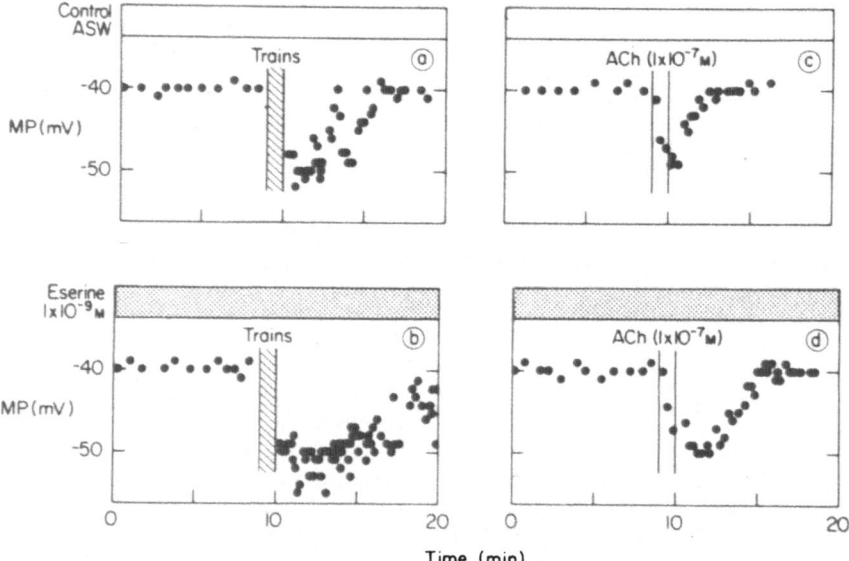

Figure 4. Effect of eserine (physostigmine) on the hyperpolarization of the Schwann cell following prolonged stimulation of the axon or external application of acetylcholine (ACh). The results shown in (*a*) and (*b*) correspond to two different nerve fibers, one immersed in control sea water (upper graph) and the other immersed in sea water containing 10^{-9} M eserine (lower graph); during the intervals indicated by the vertical striped bars stimuli were delivered to the axons at 125/sec. The results

shown in (*c*) and (*d*) correspond to a single nerve fiber immersed first in control sea water (upper graph), and then in sea water containing 10^{-9} M eserine (lower graph); during the 1 min intervals indicated by the open vertical bars the nerve fiber was exposed to acetylcholine (10^{-7} M). The Schwann cell hyperpolarizations observed in eserine (10^{-9} M) were longer than those in the control sea water solution.

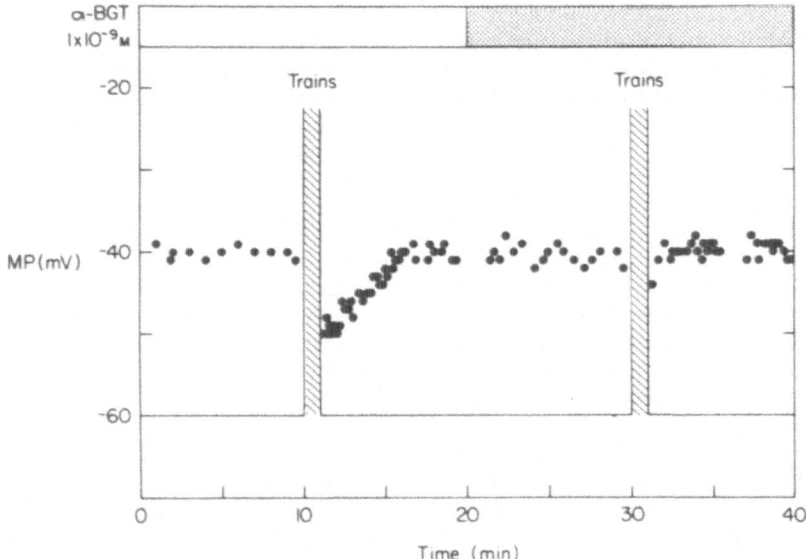

Figure 5. Effect of α-bungarotoxin (α-BGT) on the Schwann cell membrane potential and on the hyperpolarization of the Schwann cell following the conduction of nerve impulse trains by the axon. The axon membrane potential (line) was monitored. During the intervals indicated by the vertical striped bars, stimuli were delivered to the axon at 125/sec. At the time indicated by the stippled horizontal bar the toxin was added to the external sea water medium. The hyperpolarizing effect of the nerve impulse trains was abolished by the incubation in α-bungarotoxin (10^{-9} M).

Figure 5 shows that in the presence of α-BGT (10^{-9} M) the conduction of nerve impulse trains by the axon has almost no effect on the Schwann cell electrical potential of the intact nerve fiber. The resting and action potential of the axon remained apparently unchanged during the whole experimental period.

Figure 6 shows that α-BGT (10^{-9} M) in the presence of the external application of carbamylcholine (10^{-6} M) to axon-free nerve fiber sheaths has no appreciable effects on the Schwann cell electrical potential.

Figure 7 shows that after an initial period of exposure to α-BGT (10^{-9} M) in the presence of a high concentration (10^{-5} M) of D-tubocurarine, the reimmersion of the slit nerve fiber in toxin-free solutions is fol-lowed by a recovery of the long-lasting hyperpolarizing effect of carbamylcholine on the Schwann cells.

In addition, preliminary measurements of the binding of ^{125}I-α-BGT to the two different plasma membrane fractions isolated from the stellar and fin nerves of *S. sepioidea*, confirm the presence of specific receptors for acetylcholine in the Schwann cell membrane.

CONCLUSIONS

The electrophysiological studies herein described have revealed the existence of axon-Schwann cell interactions in the squid nerve fiber which, in all probability, are mediated by cholinergic mechanisms.

The effects of acetylcholine, carbamylcholine, D-tubocurarine, and α-bungarotoxin on the Schwann cell electrical potential indicate the presence of acetylcholine receptors in the plasma membrane of the Schwann cell, which is confirmed by the binding of α-BGT to the periaxonal cells plasma membrane preparations.

The effects of eserine on the long-lasting hyperpolarizations induced in the Schwann cells by the conduction of nerve impulse trains or by the external application of acetylcholine indicate that the acetylcholinesterase present in these nerve fibers may be directly involved in the mechanism responsible for the long-lasting effects of axon impulse trains on the Schwann cell.

It is still unknown whether the axon and Schwann cells in the normal intact squid nerve fiber contain, and are able to release acetylcholine as a result of nerve impulse train conduction. However, the presence of acetylcholine receptors and acetylcholinesterase enzymatic activity in the plasma membranes of these nerve fibers favors such a possibility. The experimental findings herein described confirm the existence of the axon-Schwann cell interaction previously described (15) and give support to the existence of axon-Schwann cell coupling through the acetylcholine system.

I wish to thank Drs. Carlos Sevcik and Raimundo Villegas for their critical reading of the manuscript. The help of Mr. R. Andreu Rodríguez in the drawing of the illustrations and of Mrs. Anita Froneck

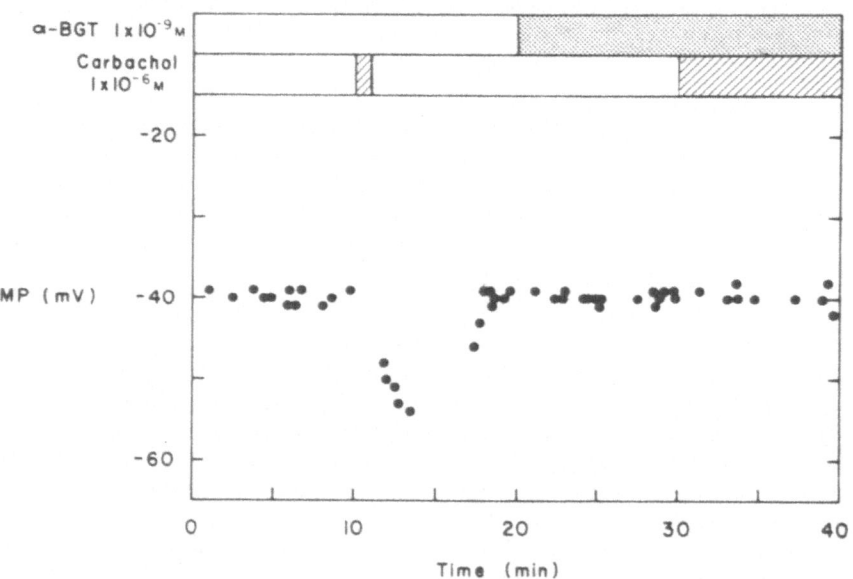

Figure 6. Effect of α-bungarotoxin (α-BGT) on the hyperpolarization of the Schwann cell following the external application of carbamylcholine to an axon-free nerve fiber sheath. During the interval indicated by the stippled horizontal bar α-BGT (10^{-9} M) was present in the bathing solution. During the intervals indicated by the striped bars the slit nerve fiber was exposed to carbachol (10^{-6} M). The hyperpolarizing effect of carbamylcholine was abolished by the incubation in α-BGT (10^{-9} M).

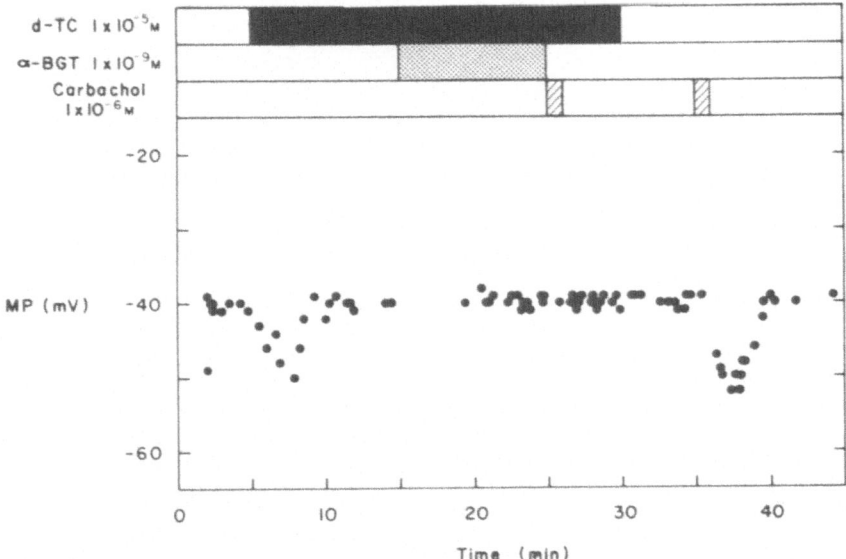

Figure 7. Changes in the membrane potential of Schwann cells in an axon-free nerve fiber sheath immersed in different test solutions. During the intervals indicated by the respective bars D-tubocurarine (shaded one), α-bungarotoxin (stippled one), and carbamylcholine (striped one) were present in the bathing solutions, at the concentrations shown in the graph. D-Tubocurarine (10^{-5} M) protects against the irreversible action of α-bungarotoxin (10^{-9} M). The reimmersion of the slit nerve fiber in toxin-free media is followed by a recovery of the hyperpolarizing effect of carbamylcholine (10^{-6} M).

Istok in the secretarial work is also gratefully acknowledged.

REFERENCES

1. CHANG, C. C., AND C. Y. LEE. *Arch. Intern. Pharmacodyn.* 144: 241, 1963.
2. CHANGEUX, J. P., M. KASAI AND C. Y. LEE. *Proc. Natl. Acad. Sci. U.S.* 67: 1241, 1970.
3. DETTBARN, W. D., AND F. A. DAVIS. *Biochim. Biophys. Acta* 66: 397, 1963.
4. FRANKENHAEUSER, B., AND A. L. HODGKIN. *J. Physiol. London* 131: 341, 1956.
5. JENKINSON, D. M. *J. Physiol. London* 152: 309, 1960.
6. KASAI, M., AND J. P. CHANGEUX. *J. Membrane Biol.* 6: 1, 1973.
7. KOENIG, E., AND G. B. KOELLE. *J. Neurochem.* 8: 169, 1961.
8. LEE, C. Y., AND C. C. CHANG. *Mem. Inst. Butantan Sao Paolo Symp. Intern.* 33: 555, 1966.
9. MILEDI, R., P. MOLINOF AND L. POTTER. *Nature* 229: 554, 1971.
10. NACHMANSOHN, D. *Chemical and Molecular Basis of Nerve Activity.* New York: Academic, 1959.
11. RITCHIE, J. M., AND C. J. ARMETT. *J. Pharmacol.* 139: 201, 1963.
12. VILLEGAS, G. M., AND J. VILLEGAS. *J. Ultrastruct. Res.* 46: 149, 1974.
13. VILLEGAS, G. M., AND R. VILLEGAS. *J. Gen. Physiol.* 51: 44s, 1968.
14. VILLEGAS, G. M., J. VILLEGAS AND P. DE WEER. *Proc. Intern. Biophys. Congr. 4th Abstr.* EIXa4/1, Moscow, 1972.
15. VILLEGAS, J. *J. Physiol. London* 225: 275, 1972.
16. VILLEGAS, J. *J. Physiol. London* 232: 193, 1973.
17. VILLEGAS, J. *J. Physiol. London* 242: 647, 1974.
18. VILLEGAS, R., L. VILLEGAS, M. GIMENEZ AND G. M. VILLEGAS. *J. Gen. Physiol.* 46: 1047, 1963.

Muscle activation: the current status

RICHARD J. PODOLSKY

National Institute of Arthritis, Metabolism and Digestive Diseases
National Institutes of Health, Bethesda, Maryland 20014

I was pleased to have been asked to organize a session in this Symposium for a number of reasons, some sentimental and some scientific. One of the sentimental ones is that Kacy Cole was one of my advisors in graduate school at the University of Chicago and I am grateful to him for the guidance he provided at that time. Kacy left Chicago before I finished my thesis work to become Technical Director of the Naval Medical Research Institute in Bethesda, but we met again when I went there to take a postdoctoral fellowship with Manuel Morales, who introduced me to muscle cells. Kacy "welcomed me aboard" when I arrived, and I began a stay that lasted a number of happy years. Kacy has been a model for many scientific workers of my generation, and his influence can be seen in our work and that of our students.

INTERNAL MEMBRANE SYSTEM

It has been known for a long time that the first step in the physiological activation of muscle cells is depolarization of the surface membrane. Considerable progress has recently been made in identifying and character-izing the subsequent processes, and it might be helpful for those of you who are not specialists in this field if I took a few minutes to describe some of these.

One of the important insights into the activation problems for striated muscle cells was the appreciation of the role played by the internal membrane system. Twenty years ago, most physiologists thought of muscle cells as nerves filled with myofibrils. However, on the basis of light microscope studies, the early histologists knew that the internal structure of the cell was far more complex than this. Figure 1, taken from a paper published by Veratti (19) in 1902, shows gold chloride-stained longitudinal and cross sections through a frog leg muscle. In the cross section, a network of fine lines surrounds the myofibrils. This transversely oriented reticulum also shows up in the longitudinal section, and it can be seen that some of the reticulum has longitudinal elements as well. This particular fiber was fixed during a local contraction; the elements of the

---

Abbreviations: SR, sarcoplasmic reticulum.

reticulum are closer together in the contracted sarcomeres than in the relaxed ones.

In frog muscle, the transverse reticulum runs through the middle of the I band, midway between successive A bands. However, in other animals, lizards, for example, the reticulum is at the junction of the A and I bands, so that there are two, rather than one, transverse lines in each sarcomere (Fig. 2). This difference was subsequently exploited by A. F. Huxley and his colleagues (9, 10), who showed with electrophysiological techniques that the transverse reticulum is the pathway along which the influence of surface membrane depolarization is conducted into the volume of the muscle fiber.

What is the nature of the reticulum? The answer to this question remained unknown until the resolving power of the electron microscope became available. Figure 3 is an osmium-fixed section of rabbit psoas muscle, published by Hugh Huxley (11) in 1957. Between the myofibrils are membrane bound vesicles, which are more frequent alongside the I band regions than in the A band regions. These internal membrane structures were much better preserved when fixation methods improved, and Fig. 4 shows what can be seen in glutaraldehyde-fixed material. The internal

membranes form triadic structures near the middle of the I bands, at the level of the Z line. Larger areas of membrane, as well as nontriadic vesicles, are seen at the A band level. Careful study of many such sections showed that the sarcoplasmic reticulum, or SR, really consists of the highly differentiated structure shown in Fig. 5, which is taken from a paper by Peachey (15). The central element of the triad at the Z line is a transverse tubule that originates at the surface membrane. The rest of the reticulum consists of longitudinally oriented cisterns, which form the outer elements of the triads, and which, in each sarcomere, are connected by longitudinal connecting tubules. Thus the internal membranes can be thought of as two systems in close apposition: the transverse tubules, and longitudinally oriented cisterns together with their connecting tubules. The two structures occupy about 15% of the cell volume (15).

The transverse tubules are open to the extracellular space. This was first shown by Hugh Huxley (12), who soaked a muscle in a ferritin solution and then examined the fibers in the electron microscope. As shown in Fig. 6, ferritin granules are present only in the central element of the triad.

This discovery, which showed that the transverse tubules are really in-

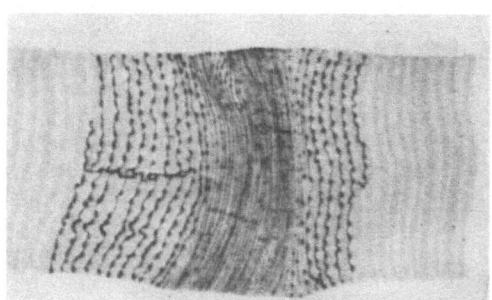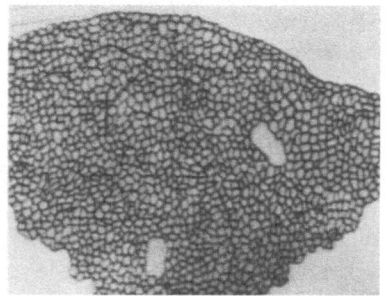

Figure 1. Sections of stained frog muscle fibers. (From Veratti (19).)

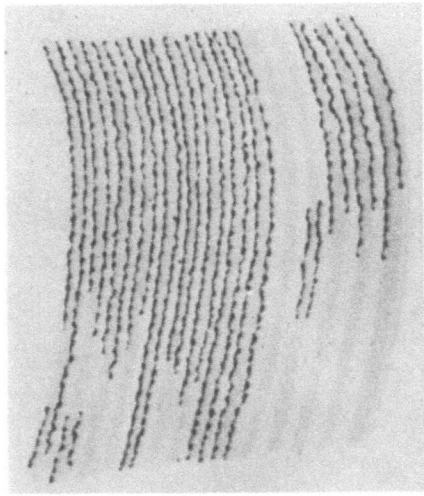

Figure 2. Longitudinal section of stained lizard muscle fiber. (From Veratti (19).)

vaginations of the surface membrane, made it useful to distinguish between the transverse tubules and the remaining elements of the internal membrane system. In what follows, I shall consider the sarcoplasmic reticulum to be those internal mem-

branes that are *not* invaginations of the surface membrane. According to this definition, the space within the SR is clearly separate from both the extracellular space and the space occupied by the myofilaments.

CALCIUM MOVEMENT

The SR contains the calcium ions that activate the myofilaments. Figure 7 shows a muscle fiber that was treated with oxalate to precipitate the calcium (1). Electron dense deposits, subsequently identified by microprobe analysis as a calcium salt (17), were found only in the terminal cisterns of the SR. This observation, together with the finding by Ebashi (3), Hasselbach and Makinose (7), and others that vesicles prepared from the internal membranes can actively accumulate calcium, led to the following picture of the activity cycle. After the action potential sweeps down the surface membrane, an electrical signal is carried inward by the transverse tubules. This effect is then communicated to the SR and,

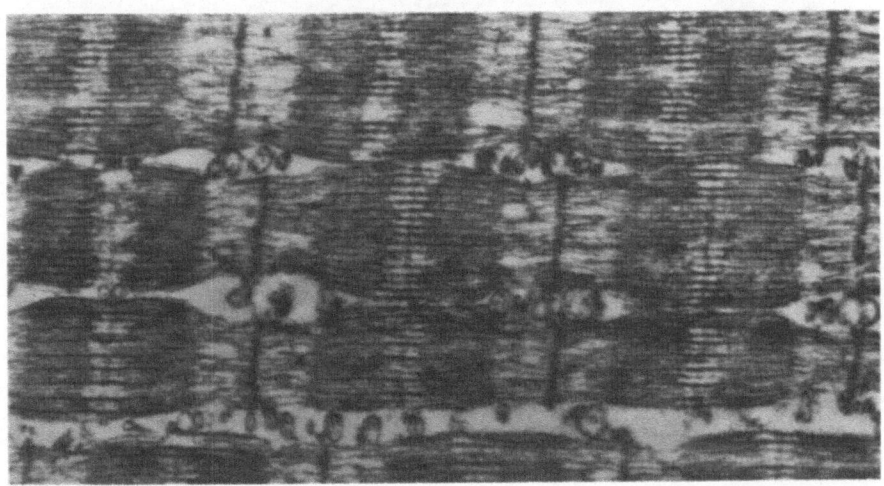

Figure 3. Electron micrograph of osmium fixed rabbit muscle fiber. (From Huxley (11).)

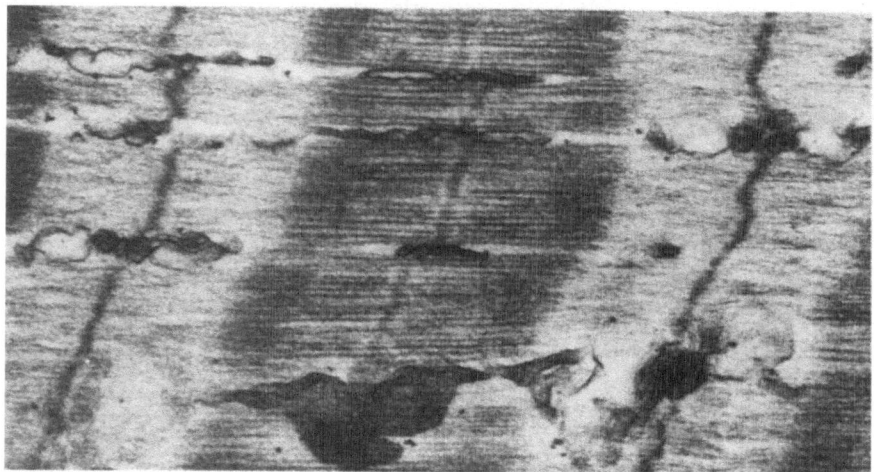

Figure 4. Electron micrograph of glutaraldehyde fixed frog muscle fiber. (From Costantin et al. (1).)

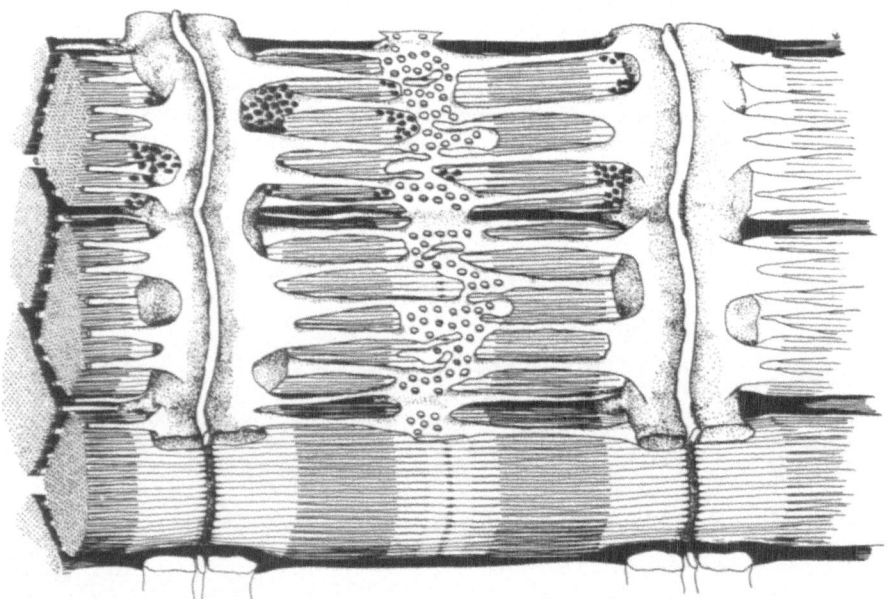

Figure 5. Reconstruction of the internal membranes associated with several myofibrils in the frog fast muscle fiber. (From Peachey (15).)

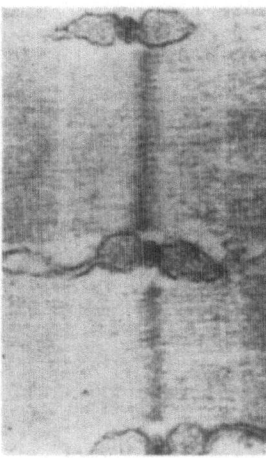

Figure 6. Electron micrograph of ferritin soaked frog muscle fiber. Note the presence of ferritin in the center element of the triad. (From Huxley (12).)

in some way, it causes the calcium to be released into the myofilament space. The released calcium binds to the myofilaments and activates the cross-bridges. At the same time calcium is pumped out of the myofilament space back into the SR, turning off the cross-bridge mechanism.

SKINNED MUSCLE FIBERS

The physiological mechanisms by which information is transferred from the transverse tubules to the SR, and by which calcium is released from the SR, are still not known. Some of the processes that might be involved have been studied in muscle fibers from which the outer membrane has been removed by microdissection. This preparation consists only of the myofilaments and the

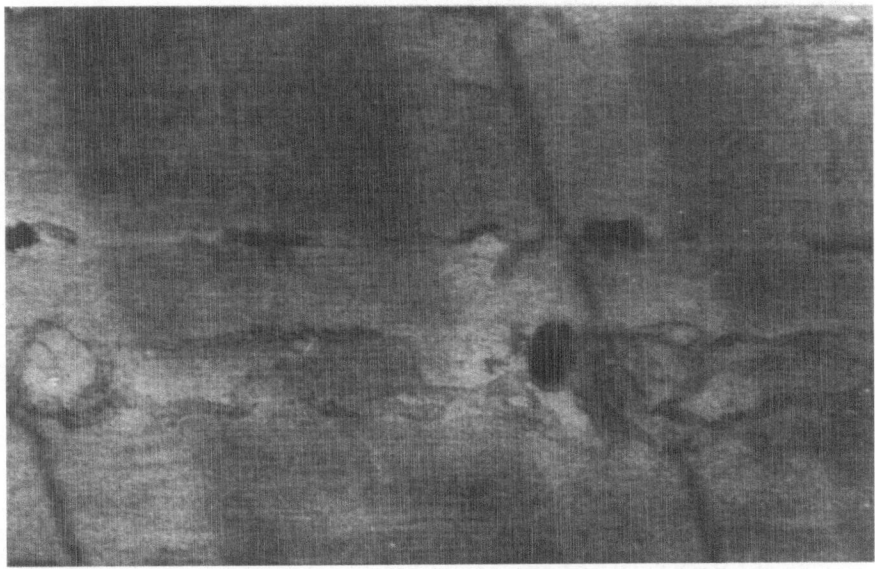

Figure 7. Oxalate treated frog muscle fiber. Note the presence of calcium oxalate in the terminal cisterns of the sarcoplasmic reticulum. (From Costantin et al. (1).)

internal membranes. The dissection technique, which is beautifully simple, was introduced exactly 20 years ago by Natori (14).

Figure 8 shows the three basic steps in our adaptation of the Natori technique. The first step, in the top panel, is to remove a small bundle of fibers from the muscle. The bundle is blotted, spread out on a glass slide, covered with mineral oil, and then, as shown in the middle panel, a single fiber is teased out. In the final step, in the lower panel, the outer membrane, generally along with a few superficial myofibrils, is dissected away. This is done with a pair of dissecting needles, one to keep the fiber from slipping, and the other to peel back the membrane. The resulting preparation is a "skinned" fiber. It can be made from muscles of many species. We have used mainly frog muscle, but skinned preparations have been prepared from fish, crayfish, horseshoe crab, and even human muscle tissue.

Originally, skinned fibers were studied with the preparations still in oil. Test solutions were applied with micropipettes and responses were scored by looking at the preparation through a microscope. Figure 9, for example, shows the response to a droplet of calcium-containing solution (16). In panel A, the calcium droplet, which is about 30 μm in diameter, is brought close to the skinned fiber. Panel B shows the initial contact. In Panel C, the sarcomeres in contact with the droplet have contracted. After a few seconds relaxation takes place, as shown in Panel D. This activity cycle can be carried out time and time again.

CALCIUM RELEASE BY DEPOLARIZATION

In addition to the direct effect of calcium, Costantin and I (2) also found that contraction occurred in certain regions of the preparation, presumably because calcium was being released from the sarcoplasmic reticulum, both when electric current was passed down the length of the fiber and when the ionic composition of the myofilament space was altered so that the internal membrane system would be expected to depolarize. We could not tell for sure which element of this system—the transverse tubules or the SR—was being depolarized; but the electric current experiments indicated that the responsive element was longitudinally, rather than transversely, oriented, which suggested that it was the SR. More recent experiments by Nakajima and Endo (13), with preparations in which the skinning technique was modified to keep the outer part of the transverse tubules from sealing off, also indicate that the SR membrane is the element that responds to depolarization.

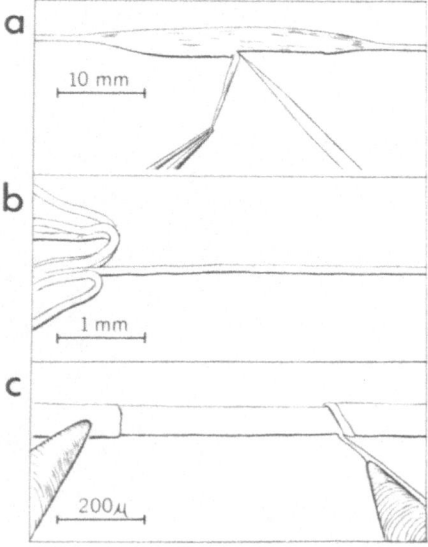

Figure 8. Technique for preparing "skinned" muscle fibers.

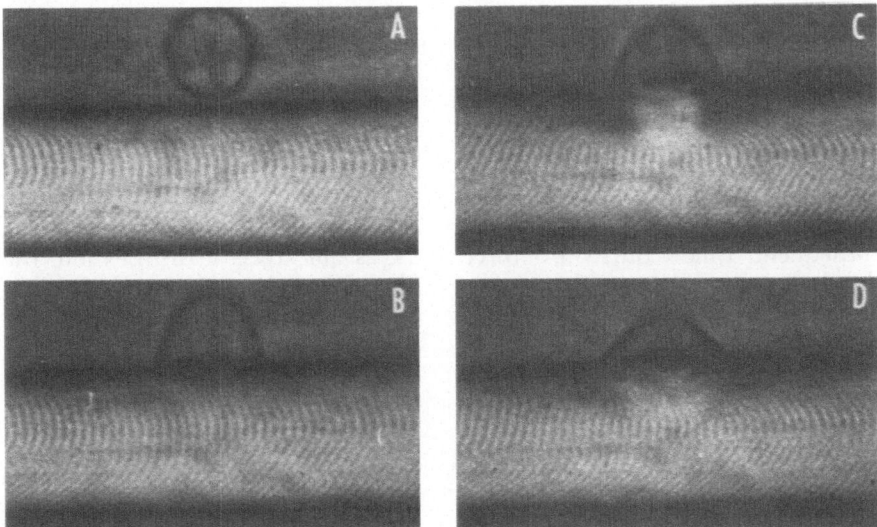

Figure 9. Response of skinned muscle fiber to droplet of calcium solution. Preparation in mineral oil. A: approach of droplet containing 3mM $CaCl_2$ + 140 mM KCl to fiber surface; B: immediately after contact of the droplet; C: 1 to 2 sec after B; D: 6 sec after B. Stage micrometer, 10 μ spacing. (From Podolsky and Costantin (16).)

One advantage of working with skinned fibers in oil is that the configuration of the internal membrane system remains normal: electron micrographs of the interior of such preparations cannot be distinguished from those of intact fibers. However, it is difficult to make quantitative measurements of the contractile response, it is difficult to control the chemical composition of the myofilament space, and it is difficult (but by no means impossible (18)) to make direct measurements of calcium movement across the SR membrane. Therefore, Hellam and I (8) worked out a way to perfuse skinned fibers with solutions of known chemical composition and to measure the contractile forces. The experimental setup is diagrammed in Fig. 10. A skinned fiber segment, several millimeters in length, is attached to two clamps. The clamp on the right is fixed, and the other is attached to a force transducer. The bathing solution can be quickly changed, and its composition can be whatever one likes.

Some data that can be obtained from such preparations are shown in Fig. 11 (4). These are records of force as a function of time, and the time scale is 30 sec. When the fiber is immersed in a potassium propionate solution containing calcium buffered with 0.5 mM EGTA to pCa 6.2, which is well above the contraction threshold, force appears only after a delay phase which lasts several minutes. During the delay phase calcium is taken up by the SR as quickly as it is released inside the fiber by the CaEGTA complex (5). After a certain amount of calcium has been accumulated by the SR, if propionate is replaced by chloride, force develops rapidly, presumably because the depolarizing effect of chloride on the SR triggers the release of calcium (Fig. 11, top panel). This explanation is supported by the second experiment in Fig. 11, which shows that addition of EGTA at the peak of force development interrupts the force spike. The third experiment shows that replacement of the chloride by propionate at the peak of the force spike does not affect the subsequent time course of force development, which would be expected if this phase of the force spike reflects the uptake of calcium by the SR. These results provide additional evidence that calcium release can be triggered when the SR is depolarized. The fourth experiment shows that the mechanical effect of released calcium is sensitive to the Mg level in the fiber.

REGENERATIVE CALCIUM RELEASE

Ford and I discovered another interesting property of the SR. This is the fact that calcium release is triggered by calcium, so that the re-

lease process is itself regenerative. Evidence for this is shown in Fig. 12 (4). The bottom trace shows the force produced by a skinned fiber immersed in a 10^{-4} M free calcium solution. Although this is about 1,000 times greater than the contraction threshold, force develops slowly; calcium is apparently taken up by the SR as quickly as it diffuses into the fiber. If, on the other hand, the fiber is exposed to free calcium after the SR has been partially loaded with calcium, a large force spike is produced at once (top trace); therefore calcium brings about the release of additional calcium. As would be expected, the spike can be interrupted by addition of EGTA (middle trace).

We checked out this interpretation of the force trace with an isotope method that measures calcium movement between the SR and the myofilament space directly. Thus, using ^{45}Ca, we showed that calcium is accumulated by the SR when the skinned fiber is exposed to an EGTA buffered calcium solution, and that

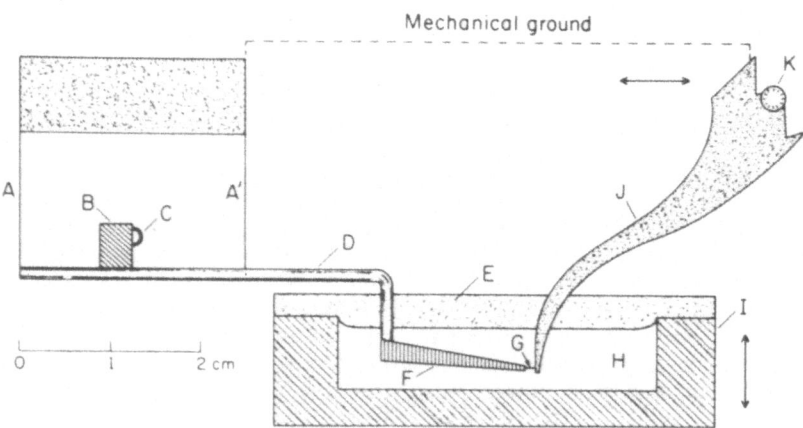

Figure 10. Perfusion chamber for recording the force development by skinned muscle fiber preparations. A, A', B, C, D, and F are parts of the force transducer; G is the fiber segment; H is the bathing solution, which is covered with oil, E, and contained in a well, I; J, K is a fixed clamp. (From Hellam and Podolsky (8).)

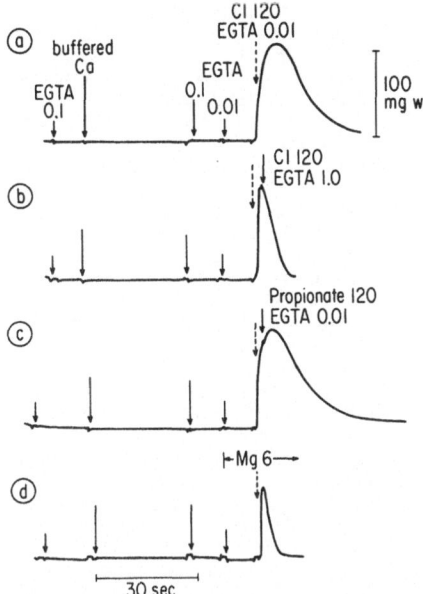

transverse tubules are more vesiculated than in normal fibers. Although the influence of these structural changes on the transport properties of the membranes remains to be investigated, it is clear that at least under certain conditions the mechanism by which calcium is released from the SR is a regenerative process.

Figure 13 indicates how these effects might be involved in physiological activation. The regenerative nature of the calcium release process in perfused skinned fibers raises the possibility that calcium acts as a transmitter of information between the transverse tubules and the SR. According to this hypothesis, the action potential would cause calcium associated with the transverse tubules to be released into the myofilament space; this "trigger" calcium would

Figure 11. Chloride-induced contractions of skinned muscle fibers. Arrows indicate solution changes; Cl 120 solutions contained 120 mM chloride in place of propionate, and Mg 6 solutions contained 6 mM $MgCl_2$ in place of 1 mM $MgCl_2$. All fibers were loaded with calcium before exposure to chloride. *a*: Quick contraction elicited by chloride; *b*: interruption of contraction by high concentration of EGTA; *c*: course of contraction not influenced by subsequent removal of chloride; *d*: attenuated contraction obtained in the presence of 6 mM magnesium (1 mM ionized magnesium plus 5 mM MgATP). (From Ford and Podolsky (4). Copyright 1970 by the American Association for the Advancement of Science.)

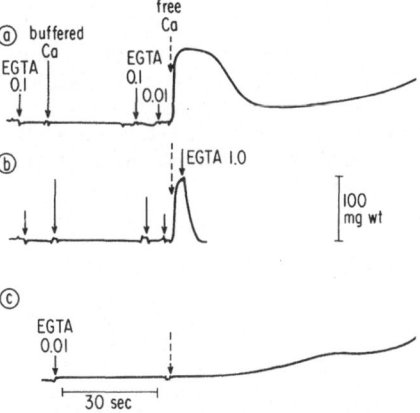

Figure 12. Calcium-induced contractions of skinned muscle fibers. Arrows mark immersion solutions containing buffered calcium, free calcium, or EGTA in the millimolar concentrations specified. *a*: Quick contraction elicited by free calcium after loading; *b*: interruption of quick contraction by high concentration of EGTA; *c*: absence of quick contraction in unloaded fiber. (From Ford and Podolsky (4). Copyright 1970 by the American Association for the Advancement of Science.)

there is *net* release of calcium when the SR is exposed to a free calcium solution(6).

We also examined the perfused skinned fibers in the electron microscope (unpublished experiments) and found that, in contrast to the situation with the fiber in oil, the structure of the internal membrane system is perturbed in many regions of the fiber: both the SR and the

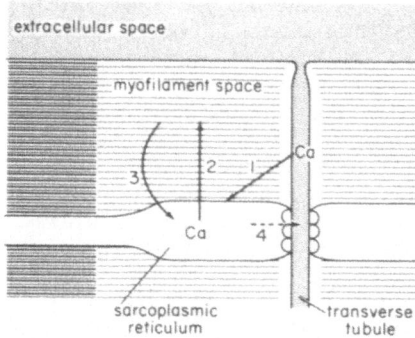

Figure 13. Possible processes in physiological activation. See text for explanation.

then elicit the release of additional calcium from the SR. An alternative hypothesis is that depolarization of the transverse tubules somehow causes the SR to depolarize and increase its calcium permeability, allowing calcium to leak out into the myofilament space. In this case, the regenerative aspect of the release process in skinned fibers might reflect an amplifying action of calcium that accelerates calcium movement out of the SR once translocation begins. These are ideas that I believe will be dealt with in more detail by the speakers in this session of the Symposium.

REFERENCES

1. COSTANTIN, L. L., C. FRANZINI-ARMSTRONG AND R. J. PODOLSKY. *Science* 147: 158, 1965.
2. COSTANTIN, L. L., AND R. J. PODOLSKY. *J. Gen. Physiol.* 50: 1101, 1967.
3. EBASHI, S. *J. Biochem. Tokyo* 50: 236, 1961.
4. FORD, L. E., AND R. J. PODOLSKY. *Science* 167: 58, 1970.
5. FORD, L. E., AND R. J. PODOLSKY. *J. Physiol. London* 223: 1, 1972.
6. FORD, L. E. AND R. J. PODOLSKY. *J. Physiol. London* 223: 21, 1972.
7. HASSELBACH, W., AND M. MAKINOSE. *Biochem. Z.* 333: 518, 1961.
8. HELLAM, D. C., AND R. J. PODOLSKY. *J. Physiol. London* 200: 807, 1969.
9. HUXLEY, A. F., AND R. W. STRAUB. *J. Physiol. London* 143: 40P, 1958.
10. HUXLEY, A. F., AND R. E. TAYLOR. *J. Physiol. London* 144: 426, 1958.
11. HUXLEY, H. E. *J. Biophys. Biochem. Cytol.* 3: 631, 1957.
12. HUXLEY, H. E. *Nature* 202: 1067, 1964.
13. NAKAJIMA, Y., AND M. ENDO. *Nature New Biol.* 246: 216, 1973.
14. NATORI, R. *Jikeikai Med. J.* 1: 119, 1954.
15. PEACHEY, L. D. *J. Cell Biol.* 25: 209, 1965.
16. PODOLSKY, R. J., AND L. L. COSTANTIN. *Federation Proc.* 23: 933, 1964.
17. PODOLSKY, R. J., T. HALL AND S. L. HATCHETT. *J. Cell Biol.* 44: 699, 1970.
18. TAYLOR, S. R., R. RÜDEL AND J. R. BLINKS. *Federation Proc.* 34: 1379, 1975.
19. VERATTI, E. *Memorie Istituto Lombardo di Scienze e Lettere.* 19: 87, Ser III, No. 10, 1902. Reprinted in: *J. Biophys. Biochem. Cytol.* 10 (No. 4, Part 2): 1, 1961.

Calcium transients in amphibian muscle[1]

S. R. TAYLOR, R. RÜDEL AND J. R. BLINKS

Department of Pharmacology, Mayo Medical School
Rochester, Minnesota 55901

ABSTRACT

The calcium-sensitive bioluminescent protein aequorin has been microinjected into isolated amphibian twitch muscle fibers in order to detect the intracellular calcium transients associated with excitation-contraction coupling. While it is not yet possible to make quantitative estimates of the changes in calcium concentration involved, it is possible to detect substantial changes in intracellular calcium transients resulting from changes in temperature, stimulation frequency, fiber length, and the osmotic strength of the bathing medium.—TAYLOR, S. R., R. RÜDEL AND J. R. BLINKS. Calcium transients in amphibian muscle. *Federation Proc.* 34: 1379–1381, 1975.

M uch has been said in the previous papers of this symposium about the role of calcium in the control of muscular activity. Although a substantial amount is already known about that role, it is probably fair to state that a detailed understanding will not be achieved until it is possible to measure changes in intracellular calcium concentration from moment to moment in a single contraction, and from place to place within a single cell. One of the most promising techniques for obtaining information of this sort involves use of the calcium-sensitive bioluminescent protein aequorin.

Aequorin, which is extracted from the luminescent jellyfish *Aequorea*

forskålea, was discovered about 12 years ago by Shimomura, Johnson, and Saiga (4) in the course of investigations into the mechanism of luminescence of the jellyfish. They found that the luminescence was due to the presence of a protein of molecular weight about 30,000 which emitted blue light in the presence of calcium ions, that no other cofactors were required, that concentra-

[1] Supported by Public Health Service grants HL 12186. and NS 10327, and by grants from the Deutsche Forschungsgemeinschaft and the Minnesota Heart Association. Use of the facilities of the Friday Harbor Laboratories, University of Washington, is gratefully acknowledged.

tions of Ca^{2+} as low as 10^{-7} M would elicit the luminescent reaction, and that the reaction appeared to be relatively specific for calcium. In 1963 (5) they suggested that aequorin might be useful for the microdetermination of calcium. This report prompted efforts in a number of laboratories, including our own, to use aequorin as a means of following calcium transients inside cells. Ashley and Ridgway (1) were the first to demonstrate the feasibility of the method, by introducing aequorin into cannulated giant muscle fibers of the barnacle. Our work has been aimed at the development of techniques for the use of aequorin as a calcium indicator in cells of more usual size, and has now progressed to the point at which we are able to record calcium transients from single amphibian skeletal muscle fibers with a fair degree of regularity. Unfortunately, the luminescent reaction of aequorin can occur only once, so great pains must be taken to protect aequorin from calcium contamination during purification and handling. The key to our technique was the production of a salt-free lyophilized preparation of aequorin in which the protein is protected from trace calcium contamination by beads of an insoluble chelating resin (Chelex 100). For experimental use, the protein is dissolved (to a concentration of about 5×10^{-5} M) in 140 mM KCl, filtered through an $0.2\ \mu$ Millipore filter, and loaded (from the butt end) into micropipettes (tip resistance 3–5 MΩ when filled with 3 M KCl). The electrical potential between the aequorin solution and the bath is monitored to verify penetration of the cell. Aequorin is injected by applying air pressure (10–100 psi) to the pipette. The volume injected is of the order of 1 nl. The injection is monitored visually through a microscope; a slight swelling of the cell can

usually be observed during a successful injection. The swelling soon subsides, and the cell appears to contract normally at the site of injection. Isometric twitches and tetani[2] are not detectably influenced by the injection.

Single twitch muscle fibers dissected from *Rana pipiens*, *Rana temporaria*, *Xenopus laevis*, and *Necturus maculosus* have been injected with aequorin in the manner just described; so far the results obtained from the various species have been similar in most essential respects. The most extensive observations have been made on fibers dissected from the tibialis anterior muscle of *R. temporaria*, and most of the records shown here were obtained from that preparation. Light was recorded with an EMI 9635B photomultiplier tube (2 inch photocathode) either mounted directly under the muscle chamber or coupled to it with a large fiberoptic probe. Force was recorded with an RCA 5734 transducer tube; temperature was maintained at 15 C unless specified otherwise.

Healthy muscle fibers injected with aequorin emit virtually no light at rest. We applied photon counting techniques to the study of this question, and found no significant increase in the dark count when a shutter was opened between the photomultiplier tube and the chamber housing the aequorin-injected muscle fiber. Changes in fiber length over a range of sarcomere spacings between 2.1 and 3.6 μm did not alter this situation. In keeping with the observation that there is no luminescence at rest is the fact that injected aequorin retains its activity for the life of the preparation (up

---

[2] This spelling of the plural of *tetanus* is used at the recommendation of the authors, although our preferred spelling, *tetanuses*, we believe is at least equally correct. Eds.

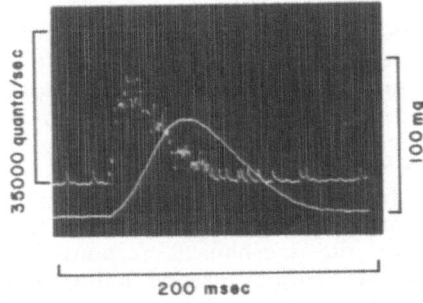

Figure 1. Luminescent and mechanical responses during a single isometric twitch. Single twitch muscle fiber dissected from tibialis anterior muscle of *R. temporaria*. Temperature 15 C; sarcomere spacing 2.4 μm. *Upper tracing*: photomultiplier output; small spikes on record are photomultiplier shot noise, and are of no biological significance. *Lower tracing*: force developed. Stimulus applied transversely through platinum field electrodes; time of stimulus is indicated by white line at bottom of record.

to 24 hours or more) unless the fiber is stimulated excessively.

A single stimulus elicits a pattern of luminescence (an "aequorin response") like that illustrated in Fig. 1. Although the details of the relationship between the luminescent and mechanical responses vary slightly from one preparation to the next, the following generalizations may be made: *1*) Force and light start to rise almost simultaneously. *2*) The aequorin response reaches its peak well before peak force is achieved. *3*) Light intensity declines to a very low level well before relaxation is complete.

In muscle fibers stimulated to twitch at infrequent intervals (>2 min), the aequorin response remains essentially unchanged over periods of many hours, indicating that a negligible proportion of the injected aequorin is consumed during a single twitch. Long periods of high frequency stimulation do lead to a

gradual exhaustion of the injected aequorin, but not at a rate that seriously interferes with the comparison of infrequent brief tetani recorded within less than an hour of one another.

Records of trains of twitches (Fig. 2) show rather striking staircase patterns that are more consistent in the case of the aequorin response than in that of the mechanical one. While peak force development usually varies in the general manner illustrated in Fig. 2, the progressive rise in force is often less striking than is shown there, and the difference between the first two beats is often more pronounced. The progressive increase in the time to peak force and in the total duration of contraction (Fig. 2, lower tracings) is consistently observed. The peak intensity of the aequorin response is regularly greatest in the first of a series of contractions, declining progressively toward a lower steady state at a rate that varies with the frequency of stimulation and with the temperature (Fig. 3). Another change that occurs during a series of twitches following a period of rest is a progressive increase in the halftime of the declining phase of the individual aequorin responses (Fig. 2). Under conditions of temperature and stimulus frequency in which summation of · the aequorin responses occurs (e.g., Fig. 3, 6 C; Fig. 5), this apparent decrease in the rate of calcium sequestration appears to manifest itself as a gradual rise in light intensity that follows the initial more rapid fall. Consistent with this interpretation is the observation that if tetani are interrupted at various times during this gradual rise of light intensity, the halftime with which the light decays becomes progressively longer as the duration of the tetanus is increased. The fact that light intensity may continue to rise long after the plateau of force is achieved is a strong indication that during a tetanus the sarcoplasmic calcium concentration considerably exceeds the level required to saturate the contractile apparatus. Other experiments (6) have indicated that this is probably true in single twitches as well, at least under some circumstances.

Changes in fiber length have a rather striking influence on the peak intensity of the aequorin response, both during single twitches (Fig. 4) and during tetani (Fig. 5). Although stretch does not markedly change the rate constant for the decline of the aequorin response in

individual twitches (Fig. 4), the gradual slowing of calcium sequestration during periods of repetitive stimulation appears to be more prominent at intermediate than at very long or very short fiber lengths (Fig. 5).

It has long been appreciated (3) that it is possible to reduce or eliminate the mechanical response of muscle without greatly influencing the action potential simply by rendering the bathing medium sufficiently hypertonic. Figure 6 shows that this effect clearly does not result from the elimination of the calcium transient. When the osmotic strength of the

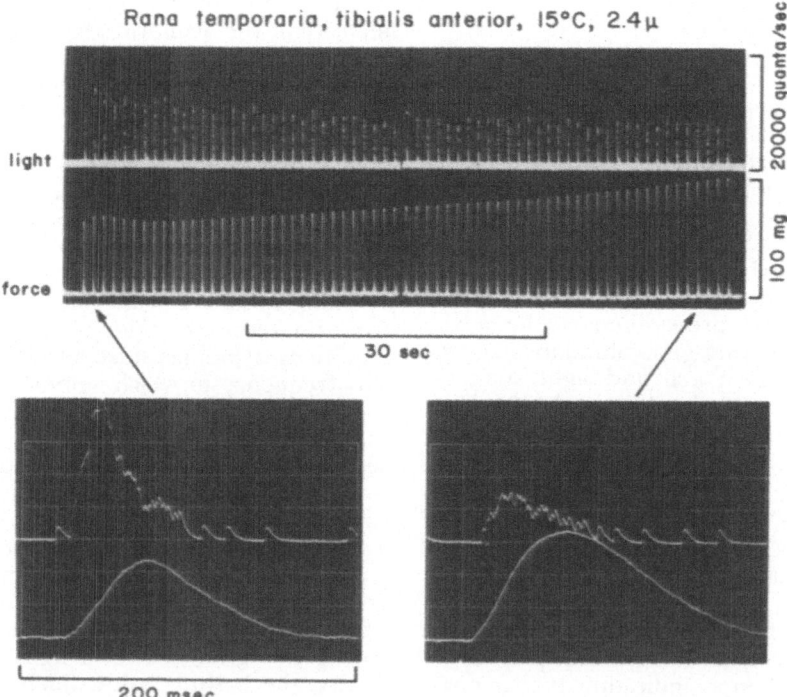

Figure 2. Staircase phenomena in luminescent and mechanical responses during a train of isometric twitches. Muscle fiber had gone unstimulated for a period of 5 min before the start of a train of stimuli applied at 1-sec intervals. Lower records are high-speed tracings of the first (*left*) and sixtieth (*right*) contractions in the series.

Rana temporaria, semitendinosus, 2.3 µ, 5H$_z$

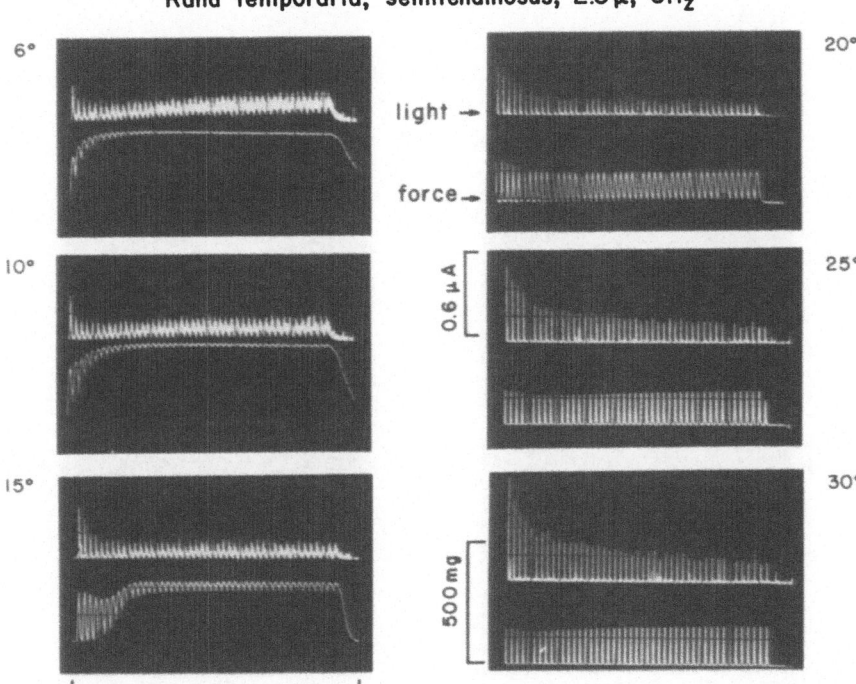

Figure 3. Influence of temperature on staircase phenomena and fusion of luminescent and mechanical responses. Each panel shows responses during a 10-sec period of 5 Hz stimulation at the temperature indicated. Fiber was unstimulated for a period of at least 5 min before the start of each tracing.

Ringer solution was doubled by the addition of sucrose, the mechanical twitch was eliminated, while the aequorin response was undiminished in amplitude and actually prolonged. In some experiments the amplitude of the aequorin response was also increased somewhat in hypertonic solutions, but the significance of this observation is clouded to some extent by the geometrical change produced by shrinkage of the fiber.

For reasons that have been spelled out elsewhere (2), it has not so far been possible for us to make quantitative estimates of the changes in calcium concentrations that are reflected in the aequorin responses that we measure. Nor are we certain, without yet having applied image intensification microscopy, of the subcellular distribution of the sites of light emission. We hope that future work will allow us to be more specific about both of these points. In the meantime, we feel that qualitative observations like those presented in this paper will continue to be capable

Rana temporaria, tibialis anterior, 15°C

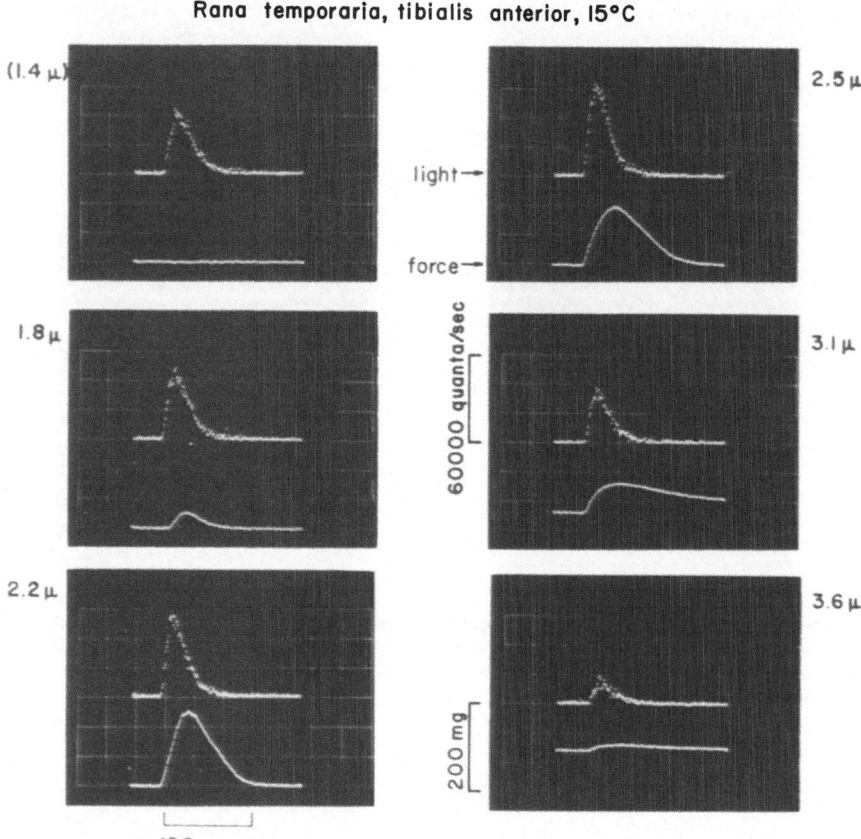

Figure 4. Influence of fiber length on luminescent and mechanical responses during isometric twitches. Muscle fiber was stimulated with single pulses at 2-min intervals. Five successive twitches were recorded at each fiber length, and averaged on a Mnemotron computer of average transients to reduce photomultiplier shot noise. Sarcomere spacing was measured in the middle of the fiber at slack length (2.2 μm); at other lengths it was estimated from the change in total fiber length. Parentheses around 1.4 μm reflect the fact that no tension was developed in that contraction. The number given is the estimated sarcomere spacing at which the fiber would have developed tension had it been able to shorten sufficiently.

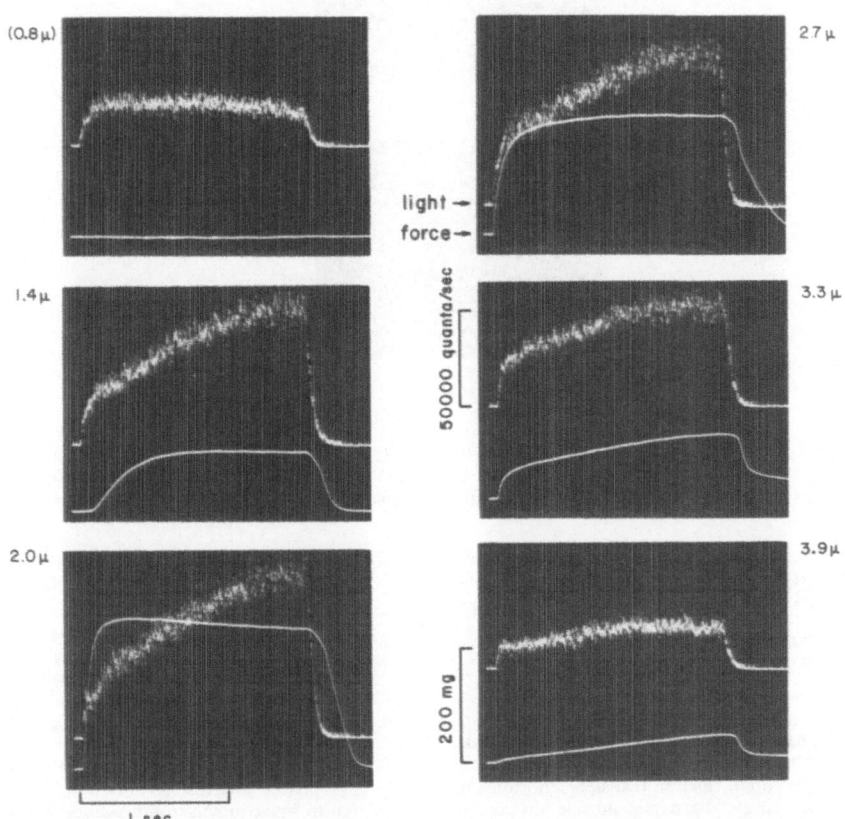

Figure 5. Influence of fiber length on luminescent and mechanical responses during isometric tetani. Baselines were adjusted in individual photographs to keep the tracings on scale. Sarcomere spacings estimated as in Fig. 4.

Rana temporaria, tibialis anterior, 15°C, 2.4 μ

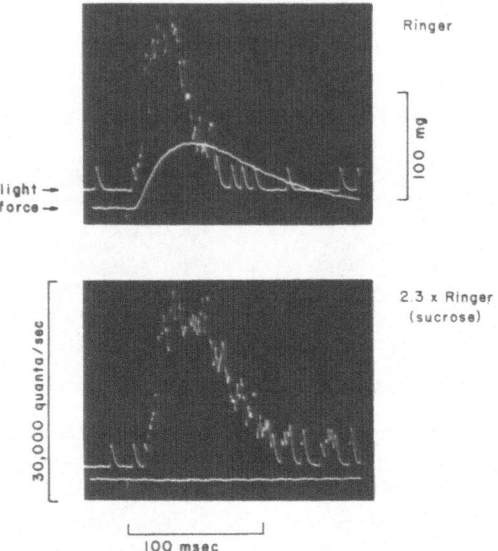

Figure 6. Influence of hypertonic bathing medium on luminescent and mechanical responses in isometric twitches. Lower panel shows response in a medium made approximately 2.3 × hypertonic by the addition of 10 g sucrose to 100 ml of Ringer solution.

of yielding a considerable amount of useful information about how muscles work.

REFERENCES

1. ASHLEY, C. C., AND E. B. RIDGWAY. On the relationships between membrane potential, calcium transient and tension in single barnacle muscle fibers. *J. Physiol. London.* 209: 105–130, 1970.
2. BLINKS, J. R. Calcium transients in striated muscle cells. *European J. Cardiol.* 1: 135–142, 1973.
3. HODGKIN, A. L., AND P. HOROWICZ. The differential action of hypertonic solutions on the twitch and action potential of a muscle fibre. *J. Physiol. London.* 136: 17–18P, 1957.
4. SHIMOMURA, O., F. H. JOHNSON AND Y. SAIGA. Extraction, purification and properties of aequorin, a bioluminescent protein from the luminous hydromedusan, *Aequorea. J. Cell. Comp. Physiol.* 59: 223–239, 1962.
5. SHIMOMURA, O., F. H. JOHNSON AND Y. SAIGA. Microdetermination of calcium by aequorin luminescence. *Science* 140: 1339–1340, 1963.
6. TAYLOR, S. R. Decreased activation in skeletal muscle fibres at short lengths. In: *The Physiological Basis of Starling's Law of the Heart.* Ciba Foundation Symposium 24 (new series). Amsterdam: Elsevier, 1974, p. 93–109.

Membrane particles and transmission at the triad[1]

CLARA FRANZINI-ARMSTRONG

Department of Physiology, University of Rochester
School of Medicine and Dentistry, Rochester, New York 14642

ABSTRACT

The structure of the membranes limiting the sarcoplasmic reticulum (SR) and transverse (T) tubules in frog and fish muscle fibers has been studied by freeze fracture. Emphasis is placed on the structure of the membranes at the triad, where thin sections have previously shown that rows of regularly disposed "feet" join SR and T tubules. Examination of the number and arrangement of particles and pits on the exposed fracture faces allows the following conclusions. 1) The SR membrane is continuous and identical in appearance along the whole sarcomere. Thus the SR is a single uninterrupted compartment and it is likely that the major function of the reticulum, calcium accumulation, is performed by the membrane limiting the lateral sacs of the triad, .as well as the longitudinal tubules. 2) At the level of the junction with the T tubule, the SR presents a strikingly different number, size and arrangement of particles and pits. This distinct portion of the SR membrane extends farther than the area covered by the junctional "feet" and no correlation can be found between the disposition of particles within the membrane and that of the feet on the membrane surface. 3) The T system membrane presents few prominent particles on its junctional face, but these are far less numerous than the feet. 4) Thus, no visible preformed channels exist between SR and T system lumina and it is suggested that direct electrical coupling between the two membranes during excitation is unlikely.—FRANZINI-ARMSTRONG, C. Membrane particles and transmission at the triad. *Federation Proc*. 34: 1382–1389, 1975.

The plasmalemma, transverse tubules and sarcoplasmic reticulum play a number of roles in the control of the activity of muscle fibers, by regulating the cycling of calcium ions. Many of the steps in excitation-contraction coupling and relaxation are well characterized, others still require clarification. Following nerve stimulation, an action potential is transmitted from end plate regions to the ends of the fiber by the plasmalemma. Depolarization spreads

[1] Supported by National Institutes of Health grant IPO LNS 1089 01-02 and by Public Health Service grant NS 08893 03.

Abbreviations: SR, Sarcoplasmic reticulum; T tubule, transverse tubule.

141

quickly to the interior of the fiber along the walls of the transverse (T) tubules, very likely by an action potential. Depolarization of the T tubules is somehow sensed by the sarcoplasmic reticulum (SR), which releases calcium previously stored in its interior. The main events following release of calcium are well established: calcium binds to troponin and removes the inhibition on actin of the troponin–tropomyosin system, thereby allowing ATP splitting and tension development (see Weber and Murray (48) for a review). Relaxation occurs as a result of calcium uptake by the SR (see Sandow (37) and Fuchs (22) for recent reviews of the pertinent literature).

A description of the membranes involved in excitation-contraction coupling thus must include two anatomically distinct membrane systems: one in contact with the extracellular space (the plasmalemma and its invaginations into the fiber, the transverse tubules), and the other forming an intracellular compartment (the sarcoplasmic reticulum). Each of these membranes has multiple functions. It turns out that areas of membranes devoted to different tasks have a distinct architecture, as seen by freeze fracture. Structural differentiations of surface and internal membranes will be described first.

Since the surface membranes and the membrane limiting the sarcoplasmic reticulum are not continuous with one another, a signal must be transmitted between them at specific junctional sites: the triads, dyads, and peripheral couplings. At triads, which are present mostly in vertebrate twitch fibers, two elements of the sarcoplasmic reticulum form a junction with one T tubule (Fig. 6). Dyads and peripheral couplings are common in myocardium, in slow fibers systems, and in the muscles of invertebrates. In these, one ele-

ment of the sarcoplasmic reticulum forms a junction either with a T tubule (in dyads), or with the plasmalemma (in peripheral couplings). It is thought that triads, dyads, and peripheral couplings are functionally analogous, since they have similar structure. Thus conclusions derived from the following description of triads in frog and fish muscle fibers are very likely applicable to dyads and peripheral couplings of other muscle fibers.

The membranes delimiting sarcoplasmic reticulum and transverse tubules are separated from each other by a narrow junctional gap. The two membrane areas facing each other across the junctional gap are named junctional SR and junctional T tubule. The rest of the SR and T tubules, which are functionally distinct from the junctional regions, are named free SR and nonjunctional T tubule respectively.

The nature of the signal crossing the junctional gap and the function of the junctional regions of SR and T tubules are yet to be established, but progress has recently been made on the steps immediately preceding and following transmission. An initial step in the coupling across the junction may be a voltage dependent movement of charge, which occurs across some part of the surface membrane, very likely the T tubules (40). On the other side of the junction, a change in the SR's membrane potential accompanies release of calcium from its interior (2). It is thought that a "depolarization" of the SR may be the step initiating calcium release (7, 30). This part of the excitation-contraction coupling process is considered in the articles by Costantin (6) and Taylor, Rüdel and Blinks (46). A fairly complete description of the architecture of the triadic junction can be obtained by comparing the images provided by freeze fracture and by conventional thin sectioning. This will constitute the second part of this paper. Hopefully, the following structural reconstruction will provide some basis for an understanding of the interactions between T tubules and SR.

The usual image obtained following freeze-fracture of tissues and cells is that of extensive exposed surfaces, belonging to membranes. These have smooth areas and are decorated by either small particles, projecting out of the surface, or small pits, penetrating into the surface. It is well documented that *1*) the fracture plane follows the hydrophobic interior of the membrane, thus exposing the inner surfaces of the two leaflets: the cytoplasmic and the exterior (or, in the case of cell organelles the luminal) leaflet (33, 47). *2*) The smooth regions of the membrane are the images of lipids in lamellar phase (9). *3*) Particles decorating the exposed fracture faces represent sites at which proteins and/or lipoprotein complexes penetrate the interior of the membrane. *4*) Pits, where visible, are the negative image of particles that have fractured on the complementary leaflet. By virtue of their size, all particles should produce a visible pit. However, this is not true, for reasons that are not well understood. The presence of pits can be used for identifying special categories of particles.

For each of the membranes to be considered here: the plasmalemma, the junctional and nonjunctional regions of the T tubules, and the junctional and free SR, the structure of the two leaflets must be examined.

Figure 1 illustrates the exposed fracture faces of the plasmalemma in twitch fibers of a small fish (*Lebistes* sp., the guppy). The circular structures, 30–40 nm in diameter, in the form of raised circles on the external leaflet and of dimples

on the cytoplasmic leaflet, are the openings of the numerous subsarcolemmal caveolae (10, 35). The external leaflet has relatively few particles, whose diameter, measured perpendicularly to the direction of shadowing, is 6–13 nm (Table 1). The cytoplasmic leaflet has a larger number of particles, also with a diameter variable between 6 and 13 nm (Table 1). In frog (*Rana pipiens*) twitch fibers, there is a more marked difference in distribution of particles between external and cytoplasmic leaflet (Table 2), but here too the larger number is on the cytoplasmic leaflet. Most cell types share with muscle fibers this characteristic of the plasmalemma.

The number of particles on the cytoplasmic leaflet of the plasmalemma in frog twitch fibers can be estimated to be between 3,000 and 6,000 per square micron. The task of defining the composition of particles in such a large heterogeneous population and of assigning each of them a function will prove very hard.

Particles in the two leaflets of the membrane of T tubules are far less numerous and more homogeneous in size than those of the plasmalemma. The transition between the particle-rich plasmalemma and the particle-poor T tubule occurs at or near the mouth of the T tubules. This is best seen in muscles of invertebrates, where the openings of T tubules are frequently included in the micrographs (21). In fish muscle fibers the nonjunctional regions of the T tubule membrane, which are visible in fractures splitting the fiber longitudinally (Fig. 2), have no particles on both cytoplasmic and luminal leaflets (Table 1, Figs. 3–5). This unusual smoothness makes the membrane of the nonjunctional regions of the T tubules similar to that of the caveolae (10). The structural similarity between the two types of invaginations of the plasma membrane is not surprising, since the T tubules initially develop as proliferations of caveolae-like structures from the periphery of the fiber to its interior (13, 39). In frog fibers the nonjunctional regions of the T tubules are somewhat different: the luminal leaflet contains numerous particles and the cytoplasmic leaflet is smooth (Table 2). This is the reverse of the disposition in the plasmalemma.

The junctional regions of the T tubules, facing towards the SR at triads, are structurally distinct from the nonjunctional regions, even though they are continuous with them. The main characteristic of the junctional T tubule is the presence of large (11–13 nm) particles, which are approximately twice as numerous on the cytoplasmic as on the luminal leaflet (Tables 1–3). The large size particles are distributed at random on the membrane (Fig. 12). Similar

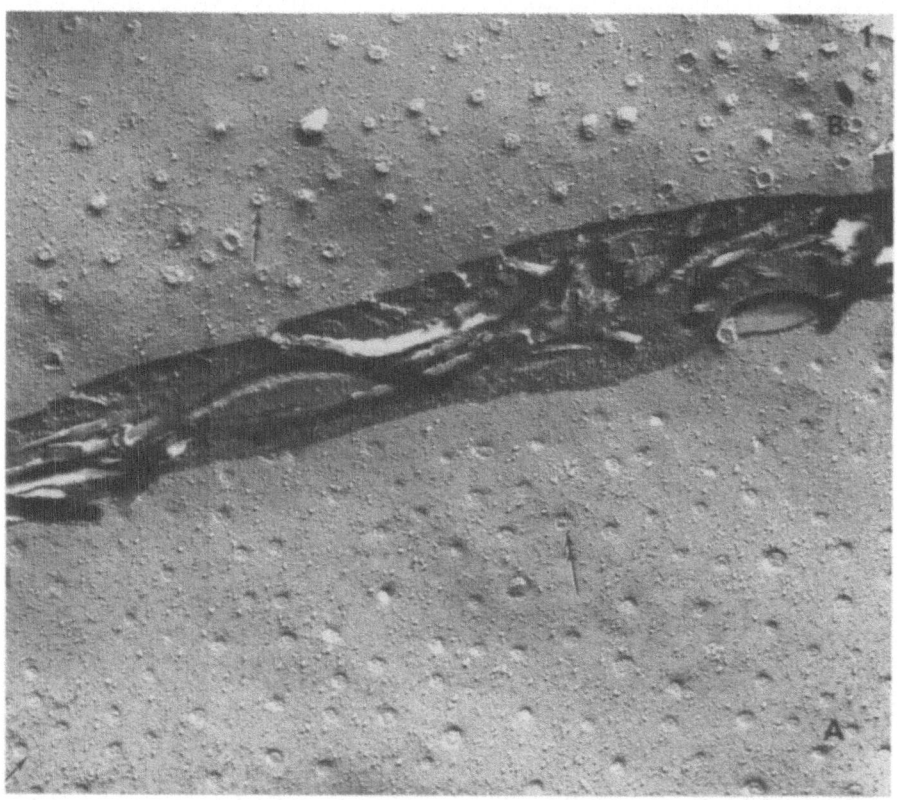

Figure 1. Exposed fracture faces of the cytoplasmic leaflet (A, below) and external leaflet (B, above) of the plasmalemma in two muscle fibers from a small fish (the guppy, *Lebistes* sp.). Double arrow: openings of the caveolae. Small bumps on both surfaces are intramembraneous particles. In this and all freeze-fracture images, black and white arrow at lower left indicates direction of shadowing. ×48,000.

TABLE 1. Distribution of particles in surface membranes of fish twitch fibers

Membrane	Leaflet	Size of particles, nm	Frequency of particles
Plasmalemma	Cytoplasmic	6–13	+++
	External	6–13	++
Nonjunctional T tubule	Cytoplasmic	—	–
	Luminal	—	–
Junctional T tubule	Cytoplasmic	12–13	++
	Luminal	12–13	+

prominent particles occupy the junctional regions of T tubules in dyads (21) and of the plasmalemma at sites of peripheral couplings (26 and unpublished observations, Peracchia and Franzini-Armstrong). The large particles that are a common component of all these junctional membranes may have a role in the process of coupling between surface membranes and SR.

The sarcoplasmic reticulum may also be divided into at least two parts: a junctional region and a free portion, which are distinct both functionally and morphologically (21). In addition, it will be here described that a third region, which is termed prejunctional SR, is interposed between free and junctional SR.

Figure 2 illustrates the path followed by the fracture plane in proximity of a triad. The fracture follows the interior of the free SR's membrane, along the longitudinal tubules (not shown) and then along the lateral sacs of the triad. Where the curvature of the membrane of the lateral sacs of the triad becomes too sharp, the fracture jumps out of the SR membrane, across the junctional gap and into the walls of the T tubules. A similar jump occurs on the other side of the T tubule, where the fracture again enters the SR. Usually, the fracture does not penetrate into the region of the junctional gap that is occupied by the feet (see below). In these fractures the free and prejunctional SR are visible, but the junctional SR cannot be seen.

The membrane of the free SR has two distinct appearances depending on whether one looks at the cytoplasmic or luminal leaflet. The cyto-

TABLE 2. Distribution of particles in surface membranes of frog twitch fibers

Membrane	Leaflet	Size of particles, nm	Frequency of particles
Plasmalemma	Cytoplasmic	4–10	++++
	External	5–10	++
Nonjunctional T tubule	Cytoplasmic	—	–
	Luminal	10–13	++
Junctional T tubule	Cytoplasmic	11–13	++
	Luminal	11–13	+

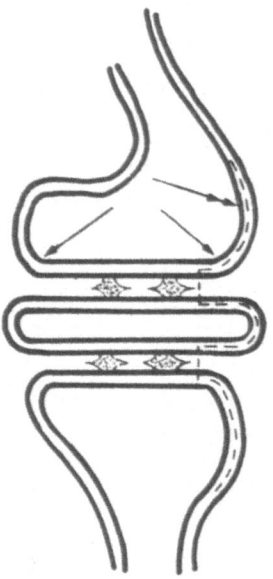

Figure 2. Diagram of a triad, showing how membranes are split by a fracture oriented parallel to the fiber's long axis. Notice that in this orientation nonjunctional regions only are visible. Double arrow shows approximate transition between free and prejunctional SR. Extent of junctional SR is indicated by two diverging arrows.

plasmic leaflet is covered by closely packed particles, approximately 8 nm in diameter. The luminal leaflet, on the other hand, is smooth. The strikingly asymmetric disposition of particles in the free SR has been confirmed in all types of muscles so far examined, including heart muscle (34, 42, 43). It has been postulated by Deamer and Baskin (8) and by Sommer et al. (43) and later confirmed by McLennan et al. (28) that the particles of the free SR represent the intramembranous, hydrophobic portion of the molecular complex constituting the calcium pump (45). The asymmetric disposition of this major protein component of the SR membrane, relative to the two leaflets, is also detectable

in the density profiles of X-ray diffraction patterns obtained from stacks of SR vesicles (11, 27).

The structure of the free SR is homogeneous, in the sense that within a given muscle fiber the density of particles on the cytoplasmic leaflet is approximately the same throughout the components of the free SR, which are, from the center of the sarcomere to one Z line (31): fenestrated collar, longitudinal cisternae, intermediate cisternae, and lateral sacs of the triad. It is likely that the pumping activity for a unit area of sarcoplasmic reticulum surface is uniform. The temporary accumulation of calcium in the SR elements at the level of the A band during relaxation (49) may very well be due to the much higher surface per unit fiber volume of the SR in that region.

In proximity of the triads the SR membrane changes dramatically in structure, to form the prejunctional SR. The change is best visible on the luminal leaflet, which, at the level indicated by a small arrow in Fig. 4, is occupied by distinct pits. Usually two rows of pits are visible, the pits of one row being displaced of a half period relative to those of the other row. A regular distance of approximately 300 Å separates the pits along each row. In the cytoplasmic leaflet, the transition between free and prejunctional SR is marked by the disappearance of the small, tightly arranged particles of the calcium pump and the appearance of larger particles, separated by smooth areas of membrane (arrowheads, Figs. 4 and 5). It is likely that the large particles of the cytoplasmic leaflet of the prejunctional SR are responsible for the pits visible on the luminal leaflet.

Notice that the structural changes between free and prejunctional SR occur in a region not facing towards the T tubules (Fig. 2). The junc-

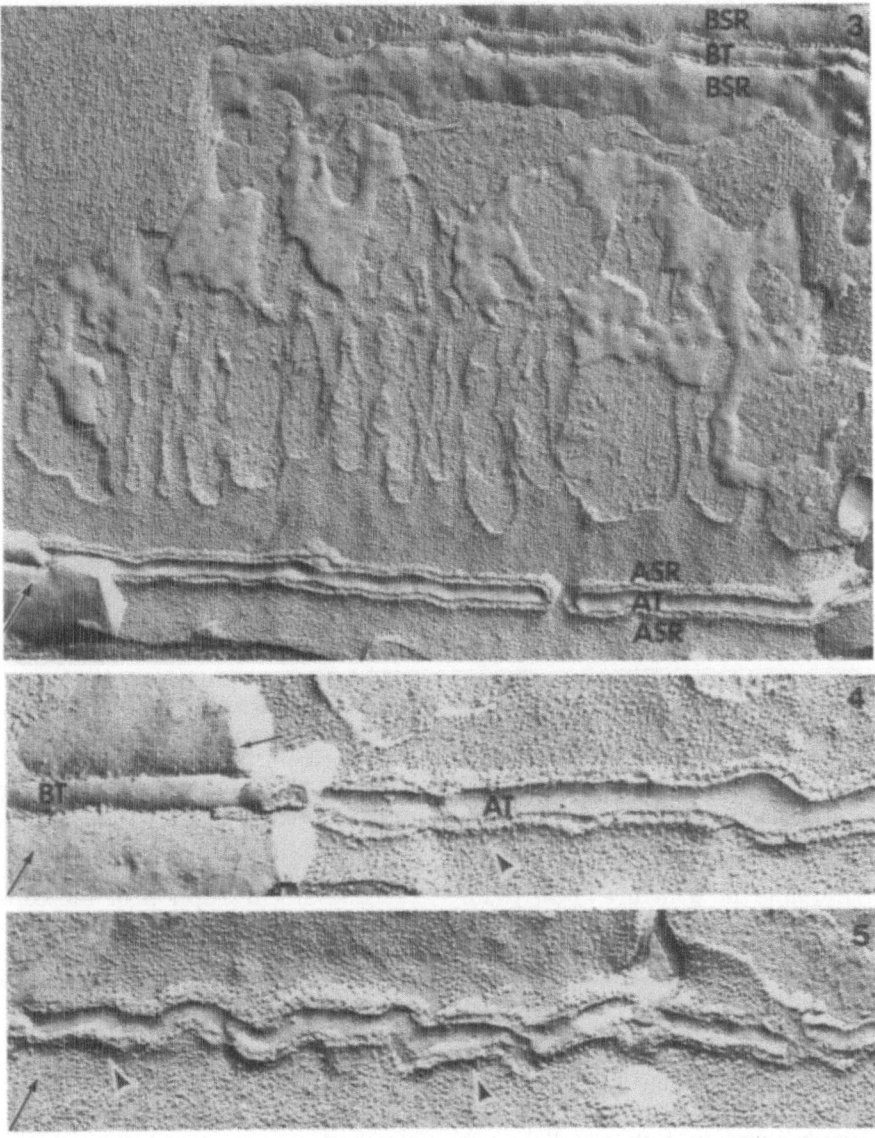

Figure 3. Freeze fracture along a segment of sarcoplasmic reticulum covering the length of one sarcomere in fish muscle. In triad at top, the luminal leaflets of SR and T tubules (BSR and BT) are shown. At bottom, the cytoplasmic leaflets (ASR and AT). Free SR has a particle rich cytoplasmic leaflet and smooth luminal leaflet. ×43,000. **Figures 4 and 5.** Details of triads. Transition between free and prejunctional SR is indicated by small arrow on luminal leaflet (Fig. 4). Arrowheads indicate transition on the cytoplasmic leaflet. Notice particle-free cytoplasmic and luminal leaflets on nonjunctional T tubules (AT and BT). ×70,000.

TABLE 3. Frequency of particles and feet
in frog triads

Membrane	Leaflet	Number of particles	Number of feet
Number/μm² of junctional T tubule surface			
Junctional	Cyto-plasmic	720	1,580
T tubule	Luminal	370	
Number/μm² of total T tubule surface			
Junctional	Cyto-plasmic	360	790
T tubule	Luminal	180	

tional SR, facing towards the T tubules, has one major characteristic in common with the prejunctional SR: its particles also leave visible pits on the luminal leaflet (see below). This is in contrast to the free SR, in which there are no pits. Thus the areas of the SR's membrane, where presumably calcium release is initiated, differ in molecular architecture from the membrane responsible for calcium pumping.

Before considering in detail the structure of the junctional regions of SR and T tubules, it is informative to examine the structure of the triad, as seen in conventional thin sectioning. The junctional SR and T tubules are separated by a gap of approximately 12 nm, which is occupied by rows of regularly arranged junctional "feet" (Fig. 6; (17)). In the triads of most twitch fibers of vertebrates, the feet form two, less frequently three or four parallel rows (Figs. 6, 8, 10; (19, 25)). Along the rows the feet are disposed at a distance of 25–30 nm (Fig. 7) and the overall disposition is tetragonal (Figs. 9, 10). When viewed in the plane of the junction, the feet have a diamond shape, their four corners

joining those of adjacent feet along the same rows and on the nearest rows (Figs. 9, 10). The result is a fenestrated sheet that lies in a plane bisecting the junctional gap, approximately halfway between SR and T tubules. The sheet is responsible for the dense line that in some views of the junction seems to bisect the gap (20). The space between the feet and, presumably, the perforations in the fenestrated sheet are accessible to solutes diffusing between the sarcoplasm and the junctional gap. Electron dense tracers, such as ferritin, introduced in the sarcoplasm of skinned fibers diffuse into at least some part of the junctional gap (18).

The disposition of feet in dyads and peripheral couplings is similar to that of triads, the only difference being the larger number of parallel rows necessary to cover the round or oval junctional areas (20, 41, 50).

The feet are attached to the SR membrane, as attested by the following observations: 1) In some muscle fibers there are areas of SR covered by feet, but not facing towards a T tubule (23, 41, 50). T tubules covered by feet, but not near any SR have not been described. 2) Following disruption of the continuity of T tubules by the so-called glycerol treatment, feet are still visible on SR surfaces originally forming a junction with T tubules. 3) At the sites of attachment of the feet, the SR membrane forms a slight scalloping (16, 24, 36). On the other hand, it is not clear whether the feet are attached to the membrane of the T tubules.

The number of feet covering a square micron of junctional SR and T tubule membrane can be calculated to be 1,580 (Table 3). When referred to the total T tubule surface (exclusive of the few longitudinal exten-

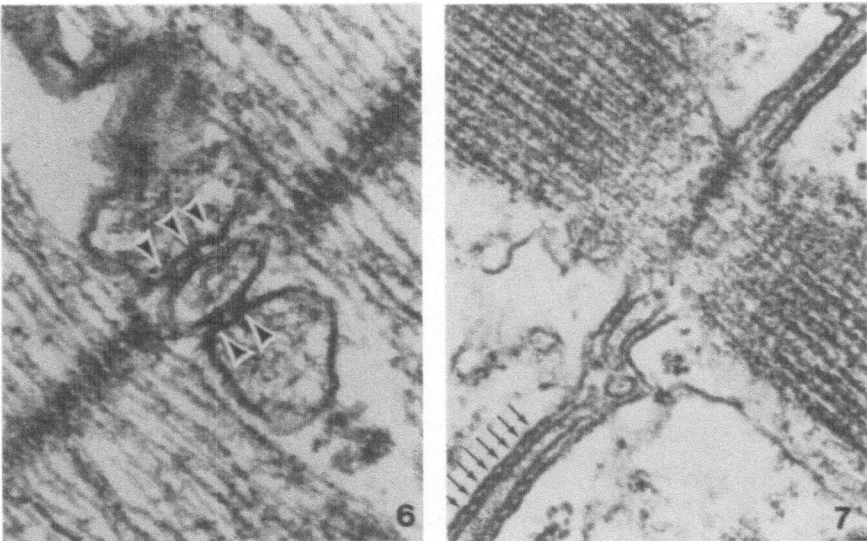

Figure 6. Section across a triad from a twitch fiber of the frog (*Rana pipiens*). In the center of the triad is a T tubule, on either side the lateral sacs of the triad, part of the sarcoplasmic reticulum. Feet (arrowheads) occupy the junctional gap between SR and T tubule membranes. ×112,000. **Figure 7.** In this view of the triad (from a frog twitch fiber) the periodical disposition of feet along a row (arrows) is visible. ×70,000.

sions) the number of feet is approximately $790/\mu m^2$.

Three possible mechanisms by which the potential across the walls of the T tubules may influence the permeability of the SR membrane to calcium have so far been proposed. One is the existence of a low resistance pathway, perhaps open only at the moment of excitation, through which ionic current may flow between the lumina of SR and T tubules (12). Since the feet are the only structure joining the two membranes, and since the junctional gap is accessible from the sarcoplasm, current flow would have to be restricted to the feet (18). One would thus expect permeability sites in the junctional membranes to be located in correspondence of the attachments of the feet, i.e., in a tetragonal arrange-

ment. A second hypothesis, formulated by Schneider and Chandler (40), proposes that an initial voltage-dependent movement of charge, across the walls of the T tubule, allows the SR membrane to sense the potential across the membrane of the T tubule by means of long molecules extending across the junctional gap. It is logical to assume that such molecules would be located in the feet. Third, coupling across the junction might involve the release of a "trigger" substance from the walls of the T tubules (Bianchi and Bolton (3)); its diffusion across the gap, probably in the space between the feet; and finally its action on the junctional SR membrane. In this case, the major function of the feet would be that of holding the junctional membranes at a fixed distance and the

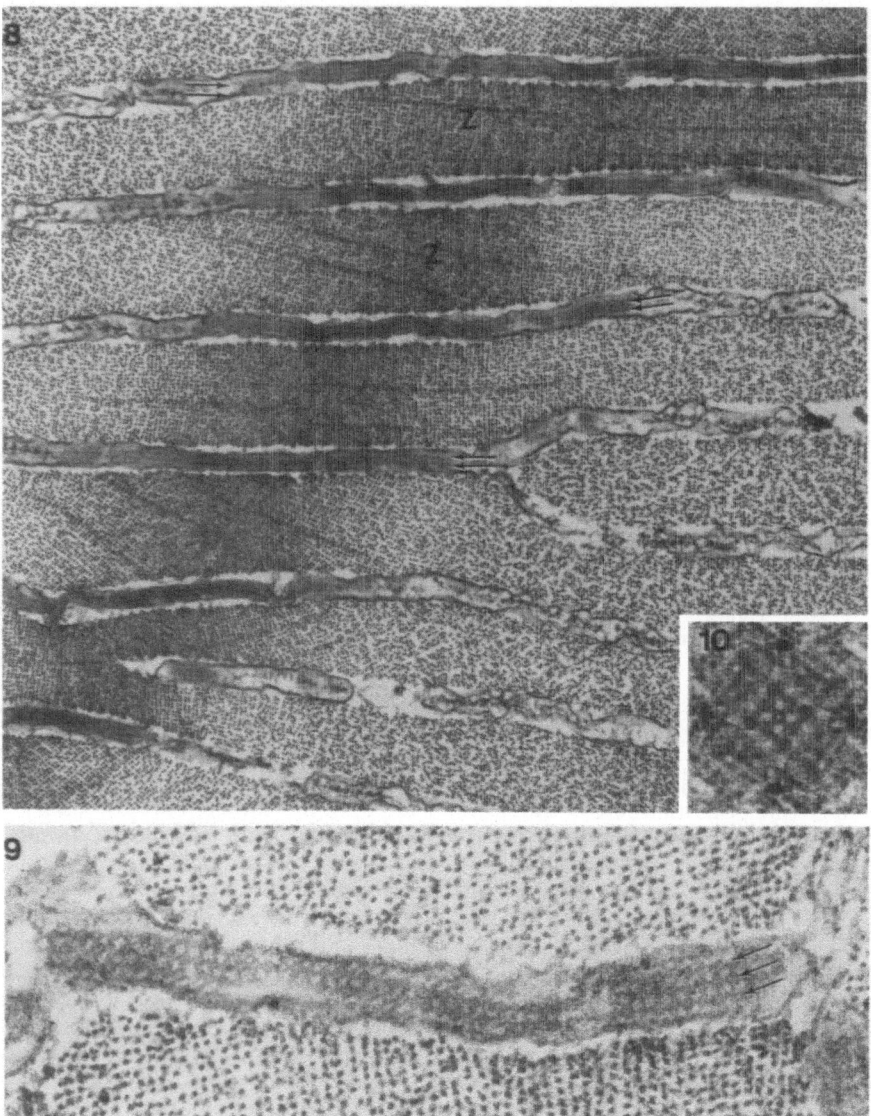

Figure 8. Cross section of a fish twitch fiber, at the level of the Z line (Z). Dark areas between the fibrils are grazing views of the junctional gap of triads. Rows of feet (usually two, arrows) occupy the gap. ×25,000. **Figure 9.** Detail of a triad in a view similar to that of Fig. 8. At right are three rows of feet (arrows). Feet have diamond shape and form a fenestrated sheet. ×70,000. **Figure 10** (inset). This photograph has been obtained by fourfold rotation centered on one junctional foot, according to Markham et al. (29). Square arrangement of feet is enhanced. ×91,000.

junctional SR membrane would be expected to have receptors on its cytoplasmic surface.

In vertebrate twitch fibers the junctional SR and T tubules are visible in fractures that break the fibers at right angle to their long axis (Figs. 11, 12). In the triads there are four membranes stacked above each other within a short distance (the two sides of a T tubule and the two junctional SR surfaces on either side). The fracture usually follows each of these membranes for a short distance and then it jumps to the nearest membrane (Figs. 11, 12). Following the direction of the arrowhead in Fig. 12, for example, the exposed fracture faces are (from right to left): a cytoplasmic leaflet of the junctional SR, luminal, and then cytoplasmic leaflet of junctional T tubule, and finally an elongated fragment of cytoplasmic T tubule, partially covering a view of the luminal leaflet of junctional SR.

Both leaflets of the junctional T tubules' membrane contain large particles, 11–13 nm in diameter, which are irregularly arranged. In frog twitch fibers the number of particles is $720/\mu m^2$ and $360/\mu m^2$ for the cytoplasmic and luminal leaflet respectively (Table 3). Notice that when referred to a unit area of junctional membrane, the number of particles on each leaflet is inferior to the number of feet (Table 3). On the other hand, the total number of particles is not significantly different from the number of feet. The particles do not match the feet in disposition, since they have no apparent symmetry of arrangement. It must be concluded that the feet are not anchored within the T tubule walls to any component of the T tubule membrane visible in freeze fracture.

The cytoplasmic leaflet of the junctional SR is also occupied by particles without a specific arrangement, but these are far more numerous than those in the junctional T tubule. As described for the prejunctional SR, the junctional SR particles leave a visible pit on the luminal leaflet (Fig. 12). The particles and pits of the junctional SR are more numerous and less regularly disposed than the feet attached to the membrane. The triad can thus be described as an "asymmetric" junction, since the dispositions of particles in the two junctional membranes do not match each other and also differ from the disposition of feet in the junctional gap. A similar asymmetry has been described in dyads of an invertebrate (21).

The lack of regular disposition of particles in SR and T tubules is somewhat disappointing. In this respect the junctional membranes at the triad differ strikingly from those forming gap junctions in epithelial cells, heart and smooth muscle, and in electrotonic synapses. At all these junctions the membranes are occupied by a regular hexagonal array of particles, matching in disposition structures that occupy the junctional gap. The particles and the matching structures in the junctional gap are thought to be the morphological basis for channels directly joining the interior of adjacent cells (see Peracchia·(32).) At the triads no preformed channels exist to provide continuity between the interior of SR and T tubules. In view of these findings, indirect evidence for openings between the lumina of SR and T tubules (4, 5) may have to be reconsidered.

The structure of the triad, as described above, is not inconsistent with an interaction between T tubules and SR of the type formulated in the "trigger" substance hypothesis. On the other hand, this hypothesis encounters two major difficulties. One

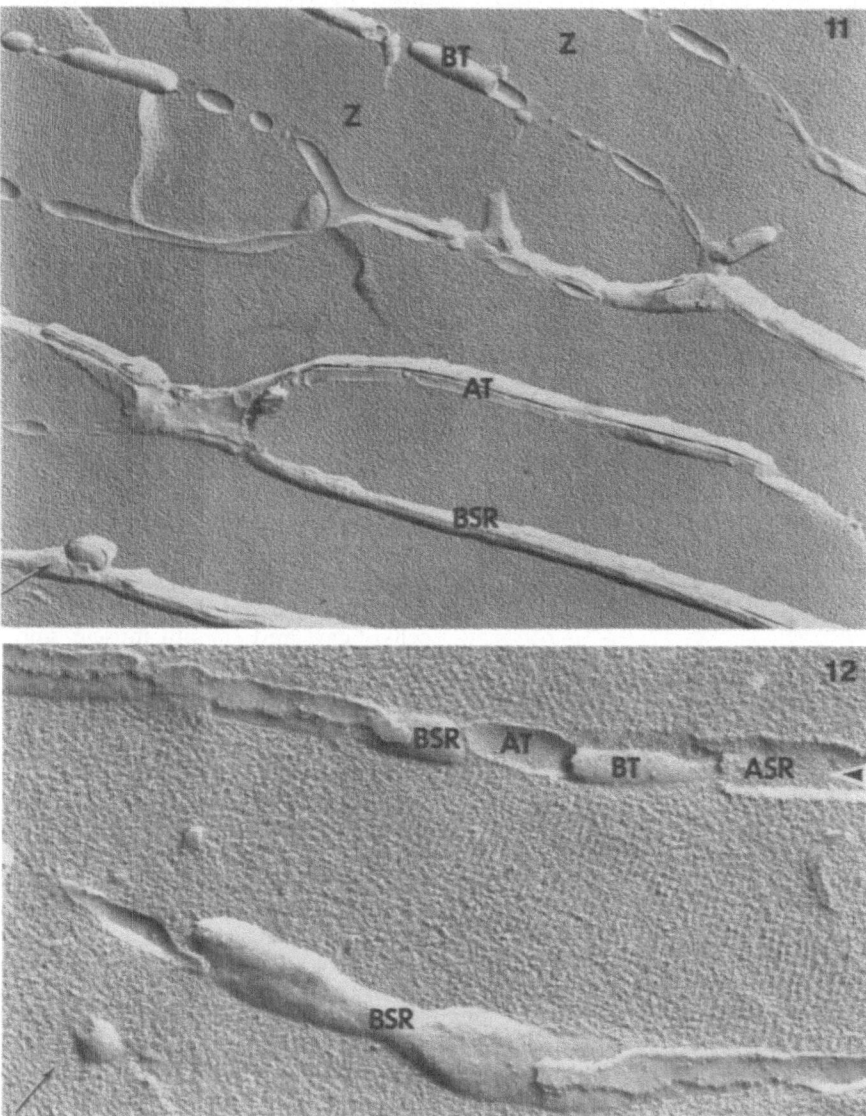

Figure 11. Freeze-fracture across a fish twitch fiber showing an area similar to that of Fig. 8. Z:Z line. Cytoplasmic and luminal leaflets of junctional T tubule (AT and BT) and luminal leaflet of junctional SR (BSR) are indicated. ×21,000. **Figure 12.** Cytoplasmic leaflet of junctional SR (ASR) has prominent particles, luminal leaflet (BSR) has pits. Compare disposition of pits with that of feet (Fig. 9). Cytoplasmic and luminal leaflet of junctional T membrane (AT and BT) have large particles, but fewer than SR. Arrowhead: see text. ×52,000.

is the nature of the trigger. Extracellular calcium entering the fiber during excitation has been ruled out by straightforward experiments, showing that twitches can be produced in the absence of extracellular calcium (1, 38). An initial fast development of tension in potassium contractures of twitch fibers is also not much affected by a reduction in extracellular calcium (44).

The second problem to be considered is the storage site for the trigger, which is nowhere to be seen. Closely related to this is the problem of synthesis and/or replenishment of the trigger at the site of release.

Finally, I would like to consider whether the e-c coupling hypothesis suggested by Schneider and Chandler (40) is consistent with the structure of the triadic junction. To do so, it is necessary to speculate on the possible role of the structures visible in the electron microscope. For example, one may assume that the intramembranous particles in the junctional T tubule are the sites of the observed capacitative charge movement and that the feet are the molecules postulated to allow interaction between the two membranes. Such assumptions are justifiable only on the basis that both the number of particles and the number of feet are not significantly different from the estimated number of charged groups, when these are referred to a unit area of T tubule surface (Table 3, see Schneider and Chandler (40)). One major difficulty remains with these assumptions and that is the nonmatching disposition of particles and feet. A more detailed knowledge of the factors affecting mobility of the particles in the T tubule membrane and of the composition of the feet is thus necessary before any conclusion can be drawn on the role of these structures in the coupling across the triadic junction.

The author thanks Mrs. Lillian Peracchia for her assistance.

REFERENCES

1. ARMSTRONG, C. M., F. M. BEZANILLA AND P. HOROWICZ. *Biochim. Biophys. Acta* 267: 605, 1972.
2. BEZANILLA, F. M., AND P. HOROWICZ. *Federation Proc.* 33: 1259, 1974.
3. BIANCHI, C. P., AND T. C. BOLTON. *J. Pharmacol. Exptl. Ther.* 157: 388, 1967.
4. BIRKS, R. I. In: *Muscle*, edited by W. H. Paul, E. E. Daniel, C. M. Kay and G. Manckton. New York: McMillan (Pergamon), 1964, p. 199–216.
5. BIRKS, R. I., AND D. F. DAVEY. *J. Physiol. London* 202: 171, 1969.
6. COSTANTIN, L. L. *Federation Proc.* 34: 1390, 1975.
7. COSTANTIN, L. L., AND R. J. PODOLSKY. *J. Gen. Physiol.* 50: 1101, 1967.
8. DEAMER, D. W., AND R. J. BASKIN. *J. Cell Biol.* 42: 296, 1969.
9. DEAMER, D. W., R. LEONARD, A. TARDIEU AND D. BRANTON. *Biochim. Biophys. Acta* 219: 47, 1970.
10. DULHUNTY, A., AND C. FRANZINI-ARMSTRONG. *Federation Proc.* 33: 401, 1974.
11. DUPONT, Y., S. C. HARRISON AND W. HASSELBACH. *Nature* 244: 555, 1973.
12. EBASHI, S., AND M. ENDO. *Progr. Biophys. Mol. Biol.* 18: 123, 1968.
13. EZERMAN, E. B., AND H. ISHIKAWA. *J. Cell Biol.* 35: 405, 1967.
14. ENDO, J., M. TANAKA AND Y. OGAWA. *Nature* 228: 34, 1970.
15. FORD, L. E., AND R. J. PODOLSKY. *Science* 167: 58, 1970.
16. FRANZINI-ARMSTRONG, C. *Federation Proc.* 23: 887, 1964.
17. FRANZINI-ARMSTRONG, C. *J. Cell Biol.* 47: 488, 1970.
18. FRANZINI-ARMSTRONG, C. *J. Cell Biol.* 49: 196, 1971.
19. FRANZINI-ARMSTRONG, C. *Tissue Cell* 4: 469, 1972.
20. FRANZINI-ARMSTRONG, C. *J. Cell Biol.* 56: 120, 1973.
21. FRANZINI-ARMSTRONG, C. *J. Cell Biol.* 61: 501, 1974.
22. FUCHS, F. *Ann. Rev. Physiol.* 36: 461, 1974.
23. JEWETT, P. H., J. R. SOMMER AND E. A. JOHNSON. *J. Cell Biol.* 49: 50, 1971.
24. KELLY, D. E. *J. Ultrastruc. Res.* 29: 37, 1969.
25. KELLY, D. E., AND J. A. CAHILL. *J. Cell Biol.* 43 (2): 66 A, 1969.
26. LANDIS, D., M. HENKART AND T. S. REESE. *J. Cell Biol.* 59 (2, Pt. 2), 184 a, 1974.

27. LIU, S. C., AND C. R. WORTHINGTON. *Biophys. Soc. Ann. Meet. Abstr.* 13: 91 a, 1973.

28. McLENNAN, D. H., P. I. SEEMAN, G. H. ILES AND G. G. YIP. *J. Biol. Chem.* 246: 2702, 1971.

29. MARKHAM, R., J. H. HITCHBORN, G. J. HILLS AND S. S. FREY. *Virology* 22: 342, 1964.

30. NAKAJIMA, Y., AND M. ENDO. *Nature* 246: 216, 1973.

31. PEACHEY, L. D. *J. Cell Biol.* 25: 209, 1965.

32. PERACCHIA, C. *J. Cell Biol.* 57: 66, 1973.

33. PINTO DA SILVA, P., AND D. BRANTON. *J. Cell Biol.* 45: 598, 1970.

34. RAYNS, D. G. *Phil. Trans. Roy. Soc. London, Ser. B Biol. Sci.* 261: 139, 1971.

35. RAYNS, D. G., F. O. SIMPSON AND W. S. BERTAUD. *J. Cell Sci.* 3: 475, 1968.

36. REVEL, J. P. *J. Cell Biol.* 12: 571, 1962.

37. SANDOW, A. *Ann. Rev. Physiol.* 32: 87, 1970.

38. SANDOW, A., M. K. D. PAGOLA AND E. C. SPHICOS. *Federation Proc.* 33: 1259, 1974.

39. SCHIAFFINO, S., AND A. MAGRETH. *J. Cell Biol.* 41: 855, 1969.

40. SCHNEIDER, M. F., AND W. K. CHANDLER. *Nature* 242: 244, 1973.

41. SHERMAN, R. G., AND A. R. LUFF. *Am. J. Zool.* 49: 1549, 1971.

42. SMITH, D. S., AND H. C. ALDRICH. *Tissue Cell* 3: 261, 1971.

43. SOMMER, J. R., R. L. STEERE, E. A. JOHNSON AND P. H. JEWETT. In: *Hibernation and Hypothermia, Perspectives and Challenges,* edited by F. F. South, J. P. Hannon, J. R. Willis, E. T. Pengelley and N. R. Alport. New York: American Elsevier, 1972, p. 291–355.

44. STEFANI, E., AND D. J. CHIARANDINI. *Arch. Ges. Physiol.* 343: 143, 1973.

45. STEWART, P. S., AND D. H. McLENNAN. *J. Biol. Chem.* 249: 985, 1974.

46. TAYLOR, S. R., R. RÜDEL AND J. R. BLINKS. *Federation Proc.* 34: 1379, 1975.

47. TILLACK, R. W., AND V. T. MARCHESI. *J. Cell Biol.* 45: 649, 1970.

48. WEBER, A., AND J. M. MURRAY. *Physiol. Rev.* 53: 612, 1973.

49. WINEGRAD, S. *J. Gen. Physiol.* 55: 77, 1970.

50. ZEBE, E., AND W. RATHMAYER. *Z. Zellforsch. Mikrosk. Anat.* 92: 377, 1970.

Electrical properties of the transverse tubular system

L. L. COSTANTIN [1]

Department of Physiology and Biophysics
Washington University School of Medicine, St. Louis, Missouri 63110

ABSTRACT

The transverse tubular system (T-system) of skeletal muscle links surface membrane action potential and release of activator calcium from sarcoplasmic reticulum (SR). The spread of depolarization along this system has been studied in voltage-clamped frog muscle fibers by using the spread of contractile activation in a thin optical cross section through the fiber center as an index. In tetrodotoxin treated fibers as depolarization of the fiber is increased contraction spreads from superficial to axial myofibrils. In tetrodotoxin free fibers the radial gradient of activation is reversed indicating that normally the activating signal is propagated along the T-system. Activation across the SR–T tubule junction does not appear to trigger an all-or-none response from the SR.—COSTANTIN, L. L. Electrical properties of the transverse tubular system. *Federation Proc.* 34: 1390–1394, 1975.

It is now generally accepted that the transverse tubular system (T-system) of skeletal muscle serves as the physiological link between the surface membrane action potential and the release of activator calcium from the sarcoplasmic reticulum (SR). This function of the T-system can be regarded as two distinct processes, *1*) transmission of the activation signal along the tubular network itself, and 2) transmission of the activation signal across the specialized junctions between individual tubular and SR elements.

Since the T-system is electrically continuous with the surface membrane, it is logical to assume that the signal that spreads along the T-system following a surface membrane action potential is simply a depolarization of the tubular network. If we ask how depolarization is transmitted along the T-system, we can suggest two possible mechanisms. The first is a regenerative depolarization of the tubular membranes, i.e., an action potential might be initiated in the

[1] Deceased Nov. 7, 1974.
Abbreviations: T-system, transverse tubular system; SR, sarcoplasmic reticulum; TTX, tetrodotoxin.

superficial tubules and propagate inward just as an action potential in the surface membrane spreads along the length of the muscle fiber. The second possibility is a passive electrotonic spread of depolarization along the T-system, the spread of depolarization by local circuit currents within the tubular network.

If we assume that the T-system is a passive cable network, we can write a set of cable equations for the T-system. The solution of these equations when a uniform steady depolarization is applied to the fiber surface is given by:

$$V(r) = V(a) \frac{I_0(r/\lambda_T)}{I_0(a/\lambda_T)} \qquad (1)$$

where

$V(r)$ is the potential across the tubular membrane at a distance r from the fiber axis.

a is the radius of the fiber, so $V(a)$ is the transmembrane potential of the tubules immediately under the surface membrane.

I_0 is the hyperbolic Bessel function of zero order.

λ_T is the length constant of the T-system.

In a one-dimensional cable an applied voltage signal decays exponentially with distance, and the rate of decay is determined by a single parameter, the length constant of the cable. Since the T-system is a radially convergent network, the solution involves, not an exponential, but a hyperbolic Bessel function. The important point, however, is that $V(r)$, the tubular membrane potential at any radial distance (r) from the fiber axis, is the function of a single parameter, λ_T, the length constant of the T-system, and just as in the more familiar one-dimensional cable, the length constant depends both on the membrane conductance and the conductance of the tubular lumen (1).

For a fiber of radius a, we can calculate the potential profile within the T-system for different values of the length constant, and this is illustrated in Fig. 1. The calculation has been done for a fiber 100 μm in diameter. The resting potential across the tubules has been taken as −90 mV, and a 35 mV steady depolarization, to a membrane potential of −55 mV, has been applied to the fiber surface. It can be seen that the passive electrotonic spread of this surface depolarization into the tubular network depends on the length constant of the T-system. When we choose a length constant comparable to the fiber radius, that is, 60 μm, there is a 5−6 mV gradient between the fiber surface and the center of the fiber; the axial T-tubules are depolarized only to about −61 mV. With a length constant twice as long, 120 μm, the gradient is <2 mV.

We cannot, of course, measure this potential profile directly; the T-tubules are simply too small to impale with microelectrodes. What we can do, however, is to use the spread of contraction to estimate the spread of potential within the T-system, and this is what Richard Adrian, Lee Peachey and I attempted to do a few years ago (2).

We started with the assumption that the contraction threshold is uniform throughout the fiber; typically the contraction threshold might be at −55.5 mV, as indicated in Fig. 1. When the surface membrane is depolarized to −55 mV, only the superficial T-tubules at either edge of the fiber will be carried beyond threshold, and the resultant contraction should involve only a superficial annulus of myofibrils. The myofibrils in the center of the fiber would not contract, because the T-tubules in this region have not been depolarized beyond threshold. When a larger depolarization is applied to the surface

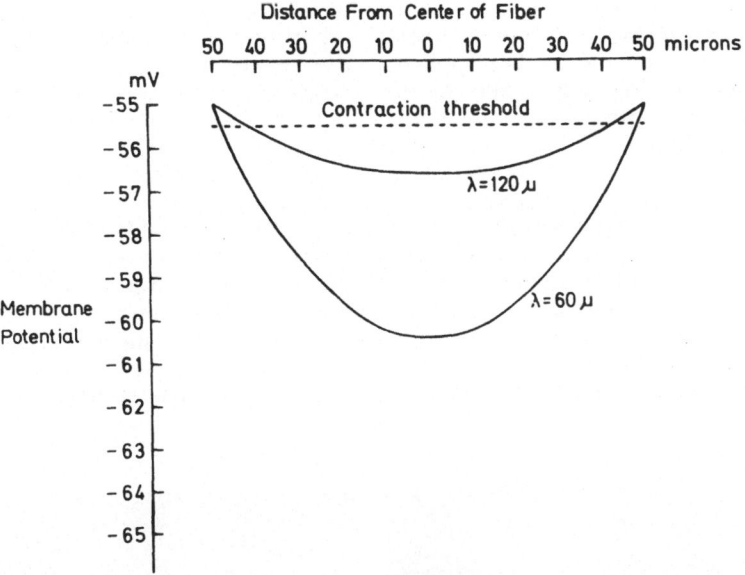

Figure 1. The steady state potential across the T-tubule membrane (*equation 1*) when the potential difference across the surface membrane of the fiber is altered from −90 to −55 mV (*V(a)* = 35 mV). The potential dis-tribution is shown for two tubular length constants (λ_T, 120 and 60 μm). The diameter of the fiber is 100 μm. The dashed line indicates the contraction threshold. (From Adrian et al. (2).)

membrane, the entire potential pro-file in the T-system will be shifted upward and, with a sufficiently large surface depolarization, the entire tubule network should be driven above threshold, so that the entire fiber cross section should contract. With this technique, then, we can use the difference between a depolariza-tion that produces a just threshold surface contraction and a depolariza-tion that activates the entire fiber cross section as an index of the poten-tial gradient within the T-system, i.e., as an index of the spread of depolari-zation along the tubular network.

Figure 2 illustrates selected frames from a cinefilm of one experiment of this type. To control the mem-brane potential and to apply step depolarizations, a voltage-clamp technique was employed. The fiber was impaled with two microelec-trodes, a voltage-recording elec-trode and, diametrically opposite, a current-passing electrode. The depth of field in the optical system was less than 20 μm, so we could obtain an optical section through the fiber. Since the plane of focus was set to pass through the center of the fiber, superficial myofibrils are seen at the two edges of the fiber in Fig. 2, and axial myofibrils are seen in the center of the fiber. This optical cross section allowed us to follow the spread of contraction inward as the surface depolarization was progressively increased.

The model calculations in Fig. 1 were done for steady state depolari-zations applied to a passive tubular network, i.e., a tubular network with-out a propagated action potential, and so the first experiment we did was to see if the real muscle fiber behaved like this model. We applied long depolarizing pulses, 200 ms in duration, and we treated the fiber with tetrodotoxin (TTX) to prevent the propagation of action potentials. When progressively larger depolarizing pulses were applied to the fiber, the first contraction we saw was immediately around the current-passing electrode. This contraction resulted from a local nonuniformity of membrane potential due to the high current density in this region, and it is of no particular interest to us. At the other side of the fiber, however, in the region of the voltage-recording electrode, the surface membrane potential was more nearly uniform. With a depolarization step just beyond threshold, as in panel B of Fig. 2, only the most superficial myofibrils in this region were contracted. The contraction did not seem to spread inward much beyond about 10 μm, and the axial sarcomeres did not shorten at all. With a depolarization step 4 mV larger, in Panel C, the contraction was more

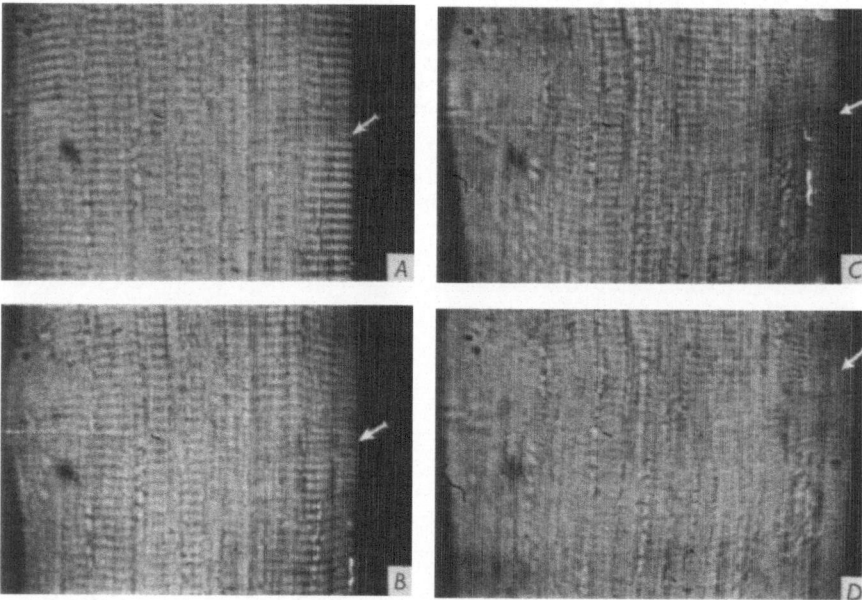

Figure 2. Inward spread of contraction in TTX-treated fiber. Frames from cinefilm of a relaxed muscle fiber (A) and of the maximum contraction elicited by 200-ms depolarizing pulses of increasing amplitude (B to D). The fiber diameter is 100 μm and the resting sarcomere length is 3.3 μm. The voltage recording electrode is at the arrow, but out of the plane of focus. The current passing electrode is opposite on the left side of the fiber; it is not clearly visualized in these records. The holding potential was -90 mV. The recorded depolarization was 39 mV (B), 43 mV (C), and 45 mV (D). In B note the contraction on the left hand edge of the fiber which appears to be restricted to the neighborhood of the current passing electrode. In B and C superficial myofibrils on the right side of the fiber are contracting. In D the axial sarcomeres are shortened to about 2.5 μm; the sarcomere length at the arrow in D is about 2.0 μm. The scale division are 10 μm. (From Adrian et al. (2).)

extensive, but a region in the center of the fiber still did not seem to contract actively. Finally in panel *D*, with a still larger depolarization, the entire fiber cross section shortened vigorously. What this experiment shows is that we can, with this technique, detect a radial gradient of activation, which presumably reflects a gradient of depolarization along the T-system. Now we are in a position to ask whether the T-system is capable of a propagated action potential. All we have to do is to leave out the TTX. If the T-system can propagate an action potential, then the depolarization signal applied to the surface membrane by the voltage clamp should spread more readily through the fiber cross section without TTX. In these experiments (5, 6), brief depolarizations (2–5 msec) were usually employed in an attempt to mimic the spread of potential along the T-system during the brief duration of a normal action potential, and a 100-msec train of pulses was applied to elicit a tetanic contraction.

The result of one of these experiments is illustrated in Fig. 3. A fiber exposed to TTX is seen in panel *A* during a vigorous contraction in response to a train of depolarizing pulses. The contraction is greatest in the superficial myofibrils while the most axial myofibrils are thrown into waves (see below). As in Fig. 2, a radial gradient of activation appears to be present. The fiber in panel *B* has not been exposed to TTX; in this fiber the contraction is greatest in the axial myofibrils while the most superficial myofibrils are thrown into waves. Thus the radial gradient of activation is reversed in the TTX-free fiber, presumably because the potential gradient along the T-system is reversed, so that the greatest depolarization now develops in the most axial T-tubules. This, of course, is just

what one would expect if the depolarizing step was large enough to activate a net inward current in the T-system. The tubules nearest the surface would be voltage-clamped at the applied depolarizing step, while the more axially located tubules would be further depolarized by the inward sodium current. In the normal muscle cell, of course, this net inward current would result in a propagated action potential along the T-system, and we can conclude that the transmission of the activation signal along the T-system normally involves the propagation of an action potential.

I want to turn now to the next step in excitation-contraction coupling in which the T-system is involved, the transmission of the activation signal across the SR-T tubule junction, and I want to ask a very simple question about how this process might work. The question is: Are we also dealing with a regenerative process here—is there a voltage threshold in the T-system that triggers an all-or-none calcium release from the SR? The alternative possibility is that tubular membrane potential controls the release of calcium from the SR in some graded fashion, so that the amount of activator made available to the contractile mechanism increases as tubular depolarization is increased. One way to approach this problem would be to change the tubular membrane potential in a graded manner and to monitor either the release of activator calcium or some index of calcium release such as the strength of contraction. As Fig. 2 illustrates, when the action potential is blocked by TTX, the contractile response of a voltage-clamped segment of a muscle fiber can be graded by grading the size of the depolarizing pulse. That is *not* what we want to examine, however. A graded contraction of this sort results from the

recruitment of more and more myofibrils with larger depolarizations; the
cross-sectional area of the fiber that is
activated is simply increased. What
we would like to test is whether
activation can be graded in a fiber
where the entire cross section is actively contracting.

To examine this problem we need
some way of knowing that the entire
fiber cross section has been activated
and that an inactive core of myofibrils
in the center of the fiber is not serving
as a load on an actively contracting
annulus of superficial myofibrils.
Fortunately, there is a way to recognize actively shortening myofibrils.
Brown, Gonzalez–Serratos, and
Huxley (3, 4) have shown that passive
myofibrils that are made to shorten
below 1.95 μm are thrown into waves,
so that if we can study contractions
in which the entire fiber cross section shortens below 1.95 μm without

waves, we will be studying actively
contracting myofibrils throughout the
fiber cross section.

These experiments were done in
collaboration with Dr. Stuart Taylor
(7), and the criterion we used for
activation of the entire fiber cross
section is illustrated in Fig. 4. The two
frames on the left are taken from a
contraction produced by a 100-ms
depolarizing pulse of 40 mV. The
fiber has shortened to 1.85 μm in
panel A and to 1.77 μm in panel B.
A wavy pattern in the axial myofibrils
can be seen in panel B, but there are
no waves in panel A. In panels C and
D, the depolarizing step was 3 mV
larger; the fiber shortened without
waves in C to 1.73 μm, while waves
appeared in panel D at a striation
spacing of 1.58 μm. In these two
contractions, then, the entire fiber
cross section was actively shortening
at least to the point shown in panel A

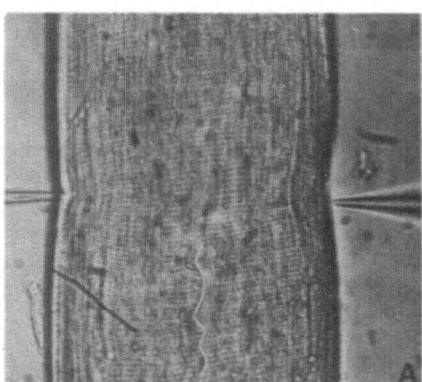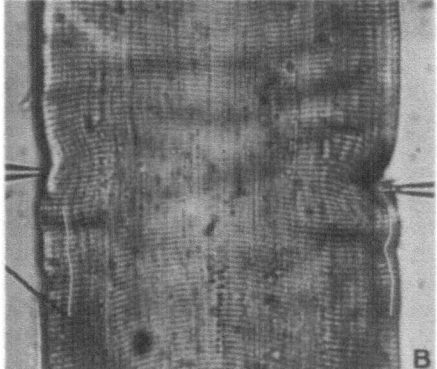

Figure 3. Frames from cine film of contractions in TTX-treated and TTX-free fibers. Holding
potential = −90 mV. Curent-passing electrode on left; voltage-recording electrode on right.
Grid spacing = 10μm. A) fiber in normal Ringer with 50 μg TTX/100 ml. Resting striation
spacing = 2.5 μm. 65-mV depolarizing pulses, 5 msec in duration. The superficial myofibrils
have shortened without waves to a striation spacing of 1.58–1.76 μm while the axial
myofibrils are thrown into waves at striation spacings of 1.65–1.79 μm. The white line traces the
pattern of the waves. B) fiber in 50% choline, 50% sodium Ringer without TTX. Resting
striation spacing = 2.5 μm. The axial myofibrils have shortened without waves to 1.58–1.77 μm,
while the superficial myofibrils are wavy at striation spacings of 1.90–2.00 μm. The white lines
trace the pattern of the waves. (From Costantin and Taylor (6).)

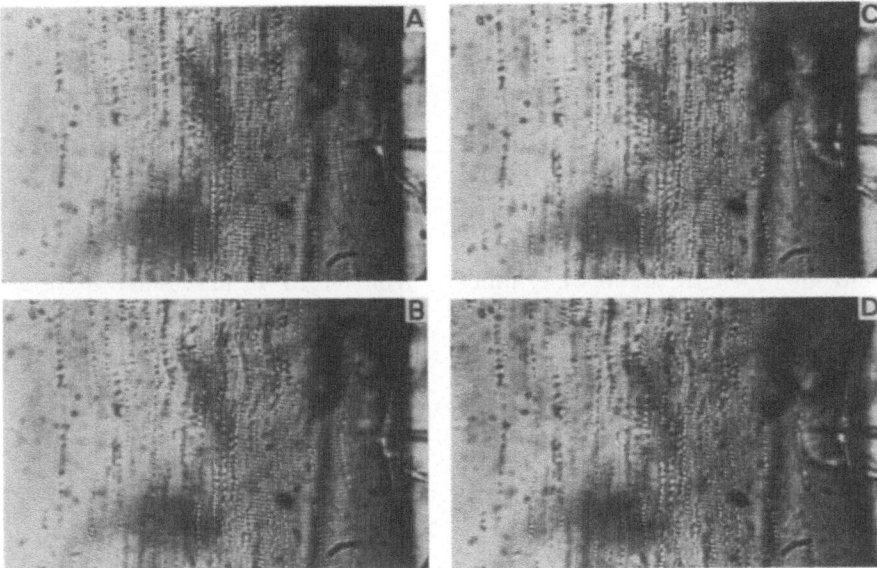

Figure 4. Selected cineframes from two successive contractions in a TTX-treated fiber. In panels *A* and *B*, contraction was elicited by a 100-msec depolarizing step of 40 mV, and in panels *C* and *D*, by a step of 43 mV. Holding potential = −90 mV. The voltage-recording electrode can be seen on the right in each panel; the tip of the current-passing electrode is just visible on the left. Striation spacing in axial myofibrils: 1.85 μm in *A*, 1.77 μm in *B*, 1.73 μm in *C*, and 1.58 μm in *D*. Note the appearance of longitudinal waves in the core of the fiber in panels *B* and *D*. Grid spacing 10 μm. (From Costantin and Taylor (7).)

and to the point shown in panel *C* with a larger depolarization. What we would like to know is whether the contraction produced by a larger depolarization is stronger. It is not possible to record the force of contraction in these fibers; the contraction involves only a short voltage-clamped segment, and the rest of the relaxed fiber in series with this segment is passively stretched by the contraction. Instead we monitored the time course of shortening in the contracting segment with high speed cine records. The time course of shortening for a series of contractions in this fiber is shown in Fig. 5. The solid lines describe shortening without waves, and further shortening after waves develop in the axial myofibrils is indicated by the interrupted

lines. In each contraction shown in the figure, the entire fiber cross section shortened to below 1.95 μm without waves, so the solid lines in each case describe the time course of active shortening of the entire fiber cross section.

Let us look at the contraction with a 40-mV depolarization. Initially the contraction is rapid; the initial slope of the shortening curve (the dotted line in Fig. 5) is quite close to the expected maximum velocity of shortening for the muscle fiber, but below about 2 μm the velocity of shortening progressively decreases. The external load on the actively contracting segment is quite small, less than 1–2% of maximum tetanic force, and the decrease in shortening velocity presumably arises from an internal re-

162   L. L. Costantin

sistance to shortening at these short sarcomere lengths. With a larger depolarization, a 43-mV depolarizing step, the fiber again begins to shorten rapidly, but in this case the rapid shortening continues to much shorter sarcomere lengths. Since the load against which this fiber is shortening is the internal resistance to shortening, the load should depend on the sarcomere length, and if we examine different contractions at the same sarcomere length, the load in each case should be the same. At any given load, a fully activated skeletal muscle should contract with a characteristic velocity. As Fig. 5 indicates, that is not what this fiber has done. At a sarcomere length of 1.9 μm, the velocity of shortening increases considerably as the depolarizing pulse is increased from 40 to 43 mV. Since there are

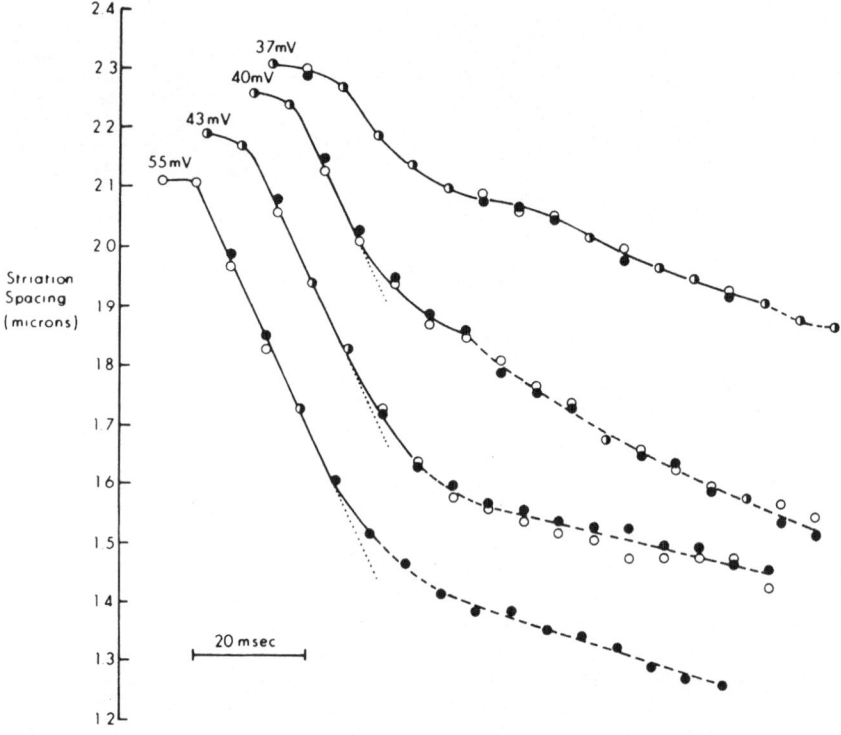

Figure 5. Time-course of shortening with increasing depolarization. Same fiber as in Fig. 4. The amplitude of the 100-msec depolarizing step is shown above each curve. The open circles represent the striation spacing of axial myofibrils measured in successive frames during each depolarization, beginning with the frame immediately before the onset of shortening. Each successive shortening curve has been displaced to the left for clarity. The filled circles are estimates of axial striation spacing obtained from measurements of extra-myofibrillar structures. Smooth curves have been fitted to the data points by eye. Solid curves indicate shortening without waves; interrupted curves indicate shortening accompanied by folding of axial myofibrils. The dotted line extending from the linear segment of the shortening curves has a slope of 20.4 μm/sec per sarcomere. (From Costantin and Taylor (7).)

no waves present, the entire fiber cross section is active in both contractions. Nevertheless, the fiber shortened further and shortened faster as the size of the depolarization step was increased. There is no sign of an all-or-none contractile response at some threshold voltage. Instead a stronger contraction can be produced even after all the myofibrils are to some degree activated, presumably because the amount of activator calcium released from the SR is increased as depolarization in the T-tubules is increased.

It should be emphasized that this sort of experiment does not provide any information about what the activation signal across the SR-T tubule junction might be. What it does indicate, however, is that the signal does not seem to trigger an all-or-none response from the SR. Instead, the T-tubule membrane potential appears to control the amount of calcium released from the SR.

REFERENCES

1. ADRIAN, R. H., W. K. CHANDLER AND A. L. HODGKIN. The kinetics of mechanical activation in frog muscle. *J. Physiol. London* 204: 207, 1969.

2. ADRIAN, R. H., L. L. COSTANTIN AND L. D. PEACHEY. Radial spread of contraction in frog muscle fibers. *J. Physiol. London* 204: 231, 1969.

3. BROWN, L. M., H. GONZALEZ-SERRATOS AND A. F. HUXLEY. Electron microscopy of muscle fibers in extreme passive shortening. *J. Physiol. London* 208: 86P, 1970.

4. BROWN, L. M., H. GONZALEZ-SERRATOS AND A. F. HUXLEY. Electron microscopy of frog muscle fibers in extreme passive shortening. *Abstr. XXV Intern. Congr. Physiol. Sci.* 243, 1971.

5. COSTANTIN, L. L. The role of sodium current in the radial spread of contraction in frog muscle fibers. *J. Gen. Physiol.* 55: 703, 1970.

6. COSTANTIN, L. L., AND S. R. TAYLOR. Active and passive shortening in voltage-clamped frog muscle fibers. *J. Physiol. London* 218: 13P, 1971.

7. COSTANTIN, L. L., AND S. R. TAYLOR. Graded activation in frog muscle fibers. *J. Gen. Physiol.* 61: 424, 1973.

Axons, dendrites and synapses

GORDON M. SHEPHERD

*Department of Physiology, Yale University School of Medicine
New Haven, Connecticut 06510*

ABSTRACT

Classical concepts of neuronal organization have been based on the motoneuron as a model. Recent work on the mitral cell of the olfactory bulb has revealed synaptic connections of dendrites and cell bodies that are not present in the motoneuron. Similar findings in many other parts of the nervous system suggest the need to revise our concepts of synaptic relations and functional properties of axons and dendrites.—SHEPHERD, G. M. Axons, dendrites and synapses. *Federation Proc.* 34: 1395–1397, 1975.

The papers at this symposium have given ample testimony to the complex properties of excitable membranes when those membranes are in the configuration of a simple cylinder, as in the giant axon of the squid. It is appropriate now to consider the properties of neuronal membranes in more complicated configurations, as in dendritic branches and in the synaptic terminals of dendrites and axons.

It is useful to have a model in mind when considering this question, and for most everyone that would be the motoneuron of the spinal cord (4). A schematic diagram of the motoneuron is shown in Fig. 1. The axon arises from the cell body and terminates on skeletal muscles in the periphery. Shorter branching processes, called dendrites, also arise from the cell body and terminate within the local gray matter of the cord. It

is known that the dendrites are the parts of the motoneuron that receive synaptic inputs and provide for spread of synaptic potentials to initiate impulses at the axon hillock; the impulses travel down the axon to activate the synapse at the neuromuscular junction onto the muscles. Thus, dendrites are the input sites, and axon terminals are the output sites, of the motoneuron. It seems such a reasonable assumption that other neurons should be organized in this way, and that the nervous system should be built up of many such input-output units.

For comparison with the motoneuron, it is useful to consider the mitral cell of the olfactory bulb. A schematic diagram of this cell is shown in Fig. 2. It can be seen that the mitral cell has a dendritic tree very similar to that of the motoneuron, save for a specialized tuft that receives the synaptic inputs from the olfactory nerves from the nose. Synaptic potentials spread through the dendrites to initiate impulses that travel through the axon to relay information to the rest of the brain.

So far this seems very similar to the arrangement of the motoneuron, but recent evidence indicates that organization of the mitral cell is much more complicated. At the first level, the mitral dendritic tuft not only receives synapses from the olfactory nerves, but also itself has synapses onto the dendrites of a type of short-axon cell, the periglomerular cell (see Fig. 2). The dendrites of the periglomerular cell, in turn, have synapses onto the mitral cell dendritic tuft (10, 22, 32, 33, 46, 47).

At a deeper level in the olfactory bulb, the mitral cell dendrites have synapses onto the dendritic spines of another type of interneuron, the granule cell, and these spines also have synapses back onto the mitral cell dendrites. These types of synapse are called reciprocal dendrodendritic synapses (23, 27, 48). Physiological (19, 21, 35, 36) and biophysical (26) studies have suggested that the mitral cell synapses are excitatory to the granule spines, and the synapses of the spines are inhibitory to the mitral cell dendrites. There is preliminary evidence at the peripheral level for similar properties: excitatory synapses from the mitral dendrites onto the periglomerular cell dendrites, and inhibitory synapses from the periglomerular dendrites onto the mitral dendrites (37).

These synaptic connections and actions provide pathways for self and lateral inhibition of mitral cells. By virtue of the large numbers of interneurons, the inhibition is very powerful, especially that mediated through the granule cells. By virtue of the fact that it is mediated largely through synaptic potentials rather than im-

pulses, the inhibition can be graded in intensity and local in character. These properties distinguish this type of inhibition from the inhibition mediated through axon collaterals and interneurons, the so-called Renshaw pathway (5, 49).

The moral of this part of the story is that most of the dendritic surface area of the mitral cell provides for synaptic output as well as synaptic input. Through these dendritic connections the mitral cell takes part in many local circuits and interactions that have no counterpart in the motoneuron.

Is this kind of organization peculiar to the olfactory bulb? Recent neuroanatomical studies suggest that it may in fact be very general. Synapses between dendrites have been found to be a prominent and probably dominant aspect of local organization in many relay nuclei of the thalamus

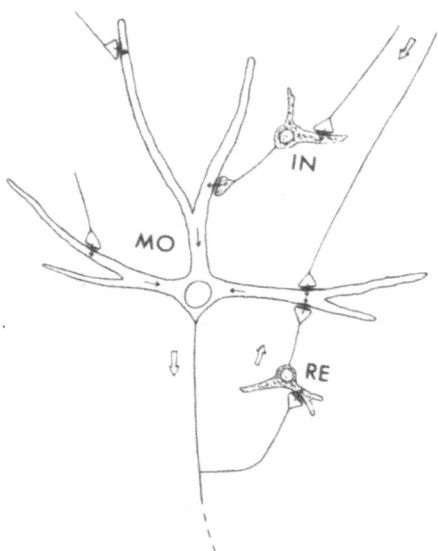

Figure 1. Schematic diagram of synaptic organization of motoneuron (MO) of spinal cord. IN: interneuron; RE: Renshaw interneuron.

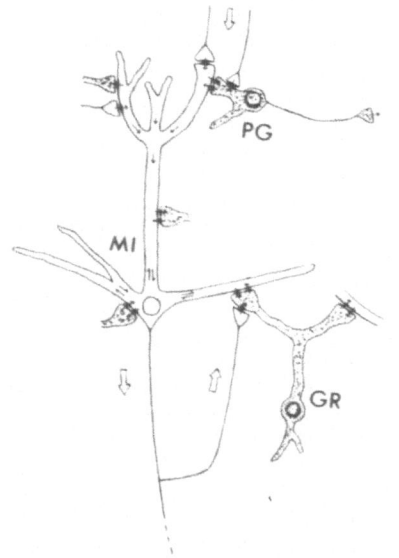

Figure 2. Schematic diagram of synaptic organization of mitral (MI) cell of mammalian olfactory bulb. PG: periglomerular short-axon cell; GR: axonless granule cell.

(7–9, 13, 14, 18, 20, 25, 34); they have also been found in the superior colliculus (16, 42), the dorsal horn of the spinal cord (28), and the motor area of the cerebral cortex of the primate (39). In the peripheral nervous system, they have been found in sympathetic ganglia (6) and the carotid body (17). Reciprocal synapses are an important aspect of the organization of the retina (3) and the carotid body (17).

These anatomical studies tell us, in effect, that our definitions of the parts of the neuron, based on the motoneuron model, are no longer valid. We cannot distinguish axon from dendrite on the basis of synaptic relation; a dendrite may send or receive synapses, and an axon terminal may receive as well as send synapses. Nor is this all: cell bodies may have synaptic outputs (e.g., 48), and synaptic inputs are found on the

initial segments and spines of axons (e.g., 45).

In response to these complications, some workers have begun to define an axon as the impulse generating process of a neuron, and a dendrite as the electrically passive part. But this can easily be shown to lack general validity. In the retina, for example, anatomists can identify an axon in bipolar and horizontal cells, yet these cells generate only graded potentials, not impulses (cf. 41, 43). On the other hand, there are by now several examples of dendrites in different parts of the nervous system that have properties of active impulse generation (15, 24, 40).

It seems wise therefore to urge the following caution. At the present stage of our knowledge it is not possible to generalize the terms axon and dendrite. The parts of the neuron can be characterized in many ways: by their outward morphology (as in Golgi-stained or dye-injected material); by the presence or absence of myelin; by characteristics of fine structure; by synaptic relations to other neurons. Physiologically, they may be characterized in terms of the ability to generate action potentials or synaptic potentials, or by their passive properties. One should probably expect that the nervous systems of vertebrates and invertebrates offer, in one place or other, virtually every possible combination of these features.

Similar remarks can be made about synapses. The classical type of synaptic terminal is the small bouton of an axon onto a motoneuron (see Fig. 1). But synaptic terminals that provide for local processing in the olfactory bulb are not at all restricted to this one type. Most of these synapses, as we have seen, provide for output from membrane that has the configuration of dendritic trunks, branches, or terminals, not to mention the synaptic outputs from cell bodies (see Fig. 2). There has been a tendency for neuroanatomists to term synapses like those onto motoneurons as "conventional," and the other types as "unconventional," "atypical," or, even, "nonusual." These terms should now be discarded. The synapses made by boutons would be better characterized as simple contacts, whereas the other types, from large terminals or membrane areas, or involving sequences or contacts, may be considered complex or specialized synapses (cf. 38). The variety of synapses—chemical and electrical—is a further aspect of the complexity of functional transmission that we can do no more than mention.

These remarks touch only briefly on recent developments in our knowledge of the organization of neurons and synapses. It seems clear that some of our classical concepts, based on the motoneuron model, have lost their generality, and that new definitions will emerge as we learn more about the synaptic relations and functional properties of axons and dendrites within the many different local circuits of the nervous system.

REFERENCES

1. ANDRES, K. H. Anatomy and ultrastructure of the olfactory bulb in fish, amphibia, reptiles, birds and mammals. In: *Ciba Foundation Symposium on Taste and Smell in Vertebrates*, edited by G. E. W. Wolstenholm and J. Knight. London: Churchill, 1970, p. 117–194.
2. BAYLOR, D. A., M. G. F. FUORTES AND P. M. O'BRYAN. Receptive fields of cones in the retina of the turtle. *J. Physiol. London* 214: 265–294, 1971.
3. DOWLING, J. E., AND B. B. BOYCOTT. Organization of the primate retina: electron microscopy. *Proc. Royal Soc. London Ser. B.* 116: 80–111, 1966.
4. ECCLES, J. C. *The Physiology of Synapses.* Berlin: Springer-Verlag, 1964.
5. ECCLES, J. C., P. FATT AND K. KOKETSU. Cholinergic and inhibitory synapses in

a pathway from motor-axon collaterals to motoneurons. *J. Physiol. London* 216: 524–562, 1955.
6. ELFVIN, L.-G. Ultrastructural studies on the synaptology of the inferior mesenteric ganglion of the cat. *J. Ultrastruct. Res.* 37: 432–448, 1971.
7. FAMIGLIETTI, E. V., JR. Dendro-dendritic synapses in the lateral geniculate of the cat. *Brain Res.* 20: 181–191, 1970.
8. FAMIGLIETTI, E. V., JR., AND A. PETERS. The synaptic glomerulus and the intrinsic neuron in the dorsal lateral geniculate nucleus of the cat. *J. Comp. Neurol.* 144: 285–334, 1972.
9. HARDING, B. M. Dendro-dendritic synapses, including reciprocal synapses, in the ventro lateral nucleus of the monkey thalamus. *Brain Res.* 34: 181–185, 1971.
10. HINDS, J. W. Reciprocal and serial dendrodendritic synapses in the glomerular layer of the rat olfactory bulb. *Brain Res.* 17: 530–534, 1970.
11. HIRATA, Y. Some observations on the fine structure of the synapses in the olfactory bulb of the mouse, with particular reference to the atypical synaptic configuration. *Arch. Histol. Japan Niigata, Japan* 24: 293–302, 1964.
12. KANEKO, A. Physiological and morphological identification of horizontal, bipolar and amacrine cells in goldfish retina. *J. Physiol. London* 207: 623–633, 1970.
13. LEVAY, S. On the neurons and synapses of the lateral geniculate nucleus of the monkey and the effects of eye enucleation. *Z. Zellforsch. Mikrosk. Anat.* 113: 396–419, 1971.
14. LIEBERMAN, A. R. Neurons with presynaptic perikarya and presynaptic dendrites in the rat geniculate nucleus. *Brain Res.* 59: 35–59, 1973.
15. LLINAS, R., AND C. NICHOLSON. Electrophysiological properties of dendrites and somata in alligator Purkinje cells. *J. Neurophysiol.* 34: 532–551, 1971.
16. LUND, R. D. Synaptic patterns of the superficial layers of the superior colliculus of the rat. *J. Comp. Neurol.* 135: 179–208, 1969.
17. MACDONALD, D., AND R. A. MITCHELL. Quantitative analysis of synaptic connections in the rat carotid body. In: *Peripheral Arterial Chemoreceptors*, edited by M. J. Purvis. Cambridge: Cambridge Univ. Press, 1975.
18. MOREST, D. K. Dendrodendritic

synapses of cells that have axons: The fine structure of the Golgi type II cell in the medial geniculate body of the cat. *Z. Anat. Entwicklungsgesch.* 133: 216–246, 1971.

19. NICOLL, R. A. Inhibitory mechanisms in the rabbit olfactory bulb: dendrodendritic mechanisms. *Brain Res.* 14: 157–172, 1969.

20. PASIK, P., T. PASIK, J. HAMORI AND J. SZENTÀGOTHAI. Golgi type II interneurons in the neuronal circuit of the monkey lateral geniculate nucleus. *Exptl. Brain Res.* 17: 18–34, 1973.

21. PHILLIPS, C. G., T. P. S. POWELL AND G. M. SHEPHERD. Responses of mitral cells to stimulation of the lateral olfactory tract in the rabbit. *J. Physiol. London* 168: 65–88, 1963.

22. PINCHING, A. J., AND T. P. S. POWELL. The neuropil of the glomeruli of the olfactory bulb. *J. Cell Sci.* 9: 347–377, 1971.

23. PRICE, J. L., AND T. P. S. POWELL. The synaptology of the granule cells of the olfactory bulb. *J. Cell Sci.* 7: 125–155, 1970.

24. PURPURA, D. P. Intracellular studies of synaptic organization in the mammalian brain. In: *Structure and Function of Synapses*, edited by G. D. Pappas and D. P. Purpura. New York: Raven, 1972, p. 257–302.

25. RAFOLS, J. A., AND F. VALVERDE. The structure of the dorsal lateral geniculate nucleus in the mouse. A golgi and electron microscopic study. *J. Comp. Neurol.* 150: 303–332, 1973.

26. RALL, W., AND G. M. SHEPHERD. Theoretical reconstruction of field potentials and dendrodendritic synaptic interactions in olfactory bulb. *J. Neurophysiol.* 31: 884–915, 1968.

27. RALL, W., G. M. SHEPHERD, T. S. REESE AND M. W. BRIGHTMAN. Dendrodendritic synaptic pathway for inhibition in the olfactory bulb. *Exptl. Neurol.* 14: 44–56, 1966.

28. RALSTON, H. J., III. The fine structure of neurons in the dorsal horn of the cat spinal cord. *J. Comp. Neurol.* 132: 275–302, 1968.

29. RALSTON, H. J., III. Evidence for presynaptic dendrites and a proposal for their mechanism of action. *Nature* 230: 585–587, 1971.

30. RALSTON, H. J., III, AND M. M. HERMAN. The fine structure of neurons and synapses in the ventrobasal thalamus of the cat. *Brain Res.* 14: 77–98, 1969.

31. REESE, T. S., AND M. W. BRIGHTMAN. Electron microscopic studies on the rat olfactory bulb. *Anat. Record* 151: 492, 1965.

32. REESE, T. S., AND M. W. BRIGHTMAN. Olfactory surface and central olfactory connections in some vertebrates. In: *Ciba Foundation Symposium on Taste and Smell in Vertebrates*, edited by G. E. W. Wolstenholme and J. Knight. London: Churchill, 1970, p. 115–149.

33. REESE, T. S., AND G. M. SHEPHERD. Dendro-dendritic synapses in the central nervous system. In: *Structure and Function of Synapses*, edited by G. D. Pappas and D. P. Purpura. New York: Raven, 1972, p. 121–136.

34. SCHEIBEL, M. E., T. L. DAVIES AND A. B. SCHEIBEL. An unusual axonless cell in the thalamus of the adult cat. *Exptl. Neurol.* 36: 512–518, 1972.

35. SHEPHERD, G. M. Responses of mitral cells to olfactory nerve volleys in the rabbit. *J. Physiol. London* 168: 89–100, 1963.

36. SHEPHERD, G. M. Neuronal systems controlling mitral cell excitability. *J. Physiol. London* 168: 101–117, 1963.

37. SHEPHERD, G. M. Physiological evidence for dendrodendritic synaptic interactions in the rabbit's olfactory glomerulus. *Brain Res.* 32: 212–217, 1971.

38. SHEPHERD, G. M. *The Synaptic Organization of the Brain*. New York: Oxford Univ. Press, 1974.

39. SLOPER, J. J. Dendro-dendritic synapses in the primate motor cortex. *Brain Res.* 34: 186–192, 1971.

40. SPENCER, W. A., AND E. R. KANDEL. Electrophysiology of hippocampal neurons. IV. Fast prepotentials. *J. Neurophysiol.* 24: 272–285, 1961.

41. STELL, W. K. The morphological organization of the vertebrate retina. In: *Handbook of Sensory Physiology. VII/1B. Physiology of Photoreceptor Organs*, edited by M. G. F. Fuortes. Berlin: Springer-Verlag. 1972, p. 111–214.

42. STERLING, P. Receptive fields and synaptic organization of the superficial gray layer of the cat superior colliculus. *Vision Res. Suppl.* 3: 309–328, 1971.

43. TOMITA, T. Electrophysiological study of the mechanism subserving color coding in the fish retina. *Cold Spring Harbor Symp. Quant. Biol.* 30: 559–566, 1965.

44. WERBLIN, F. S., AND J. D. DOWLING. Organization of the retina of the mudpuppy *Necturus maculosus*. II.

Intracellular recording. *J. Neurophysiol.* 32: 339–355, 1969.

45. WESTRUM, L. E. Observations on initial segments of axons in the prepyriform cortex of the rat. *J. Comp. Neurol.* 139: 337–356, 1970.

46. WHITE, E. L. Synaptic organization in the olfactory glomerulus of the mouse. *Brain Res.* 37: 69–80, 1972.

47. WHITE, E. L. Synaptic organization of the mammalian olfactory glomerulus: new findings including an intraspecific variation. *Brain Res.* 60: 299–313, 1973.

48. WILLEY, T. J. The ultrastructure of the cat olfactory bulb. *J. Comp. Neurol.* 152: 211–232, 1973.

49. WILLIS, W. D. The case for the Renshaw cell. *Brain Behav. Evol.* 4: 5–52, 1971.

50. YAMAMOTO, C., T. YAMAMOTO AND K. IWAMA. The inhibitory system in the olfactory bulb studied by intracellular recording. *J. Neurophysiol.* 26: 403–415, 1963.

Motoneuron dendrites: role in synaptic integration[1]

JOHN N. BARRETT

Division of Neurobiology, Department of Physiology and Biophysics
University of Iowa, Oakdale, Iowa 52319 and
Department of Physiology and Biophysics, University of Miami
Miami, Florida 33152

ABSTRACT

Dendrites constitute over 80% of the receptive surface area in cat motoneurons. Calculations based on matched electrical and geometrical measurements in these neurons indicate that the specific resistance of dendritic membranes in resting motoneurons is at least 2,000 ohm-cm². When the specific membrane resistance is this high, even the most distal dendritic synapses can contribute significantly to the depolarization of the soma, and hence influence the rate of action potential generation. However, dendritic membrane resistance depends strongly on the level of background synaptic activity. The conductance changes associated with excitatory synaptic activity on a dendrite can be great enough to reduce significantly both the excitatory synaptic driving potential and the effective membrane resistance on that dendrite, and thus greatly reduce the effectiveness of synapses on that dendrite. Inhibitory synaptic activity produces an even greater reduction in dendritic membrane resistance. Thus the relative effectiveness of dendritic synapses depends on the type, distribution, and intensity of background synaptic activity, as well as on dendritic geometry and resting membrane properties. —BARRETT, J. N. Motoneuron dendrites: role in synaptic integration. *Federation Proc.* 34: 1398–1407, 1975.

The relationship between the synaptic input to a neuron and its action potential output is basic to nervous system function. This input-output relationship is determined by the membrane properties and the geometry of the neuron. This discussion will deal primarily with synaptic integration in cat alpha motoneurons, whose electrophysiology and morphology have received extensive study. Input-output relationships differ among the various neuronal types, but many of the basic principles and mechanisms uncovered in studying motoneurons will also be useful in

[1] Supported by Public Health Service training grant NS 05748 from the National Institute of Neurological Diseases and Stroke.

understanding synaptic integration in other neurons.

Motoneurons have an extensive dendritic tree, which constitutes at least 80% of the receptive surface of the neuron. Initial estimates of dendritic properties led to the hypothesis that only those dendritic synapses close to the cell body would have any significant effect on motoneuronal function. Recent measurements and calculations, however, predict a significant functional role for all dendritic synapses in the resting neuron, but show that the effectiveness of dendritic synapses will be very strongly influenced by background synaptic activity. These recent studies will be described below, after a brief introduction to the synaptic physiology of motoneurons (see also 20, 31).

Synaptic currents

Afferent fibers from many different sources form synapses on the dendrites and soma (cell body) of motoneurons. The morphology and physiology of these synapses indicate that they are chemically mediated. Activated synaptic terminals release packets (or quanta) of transmitter substances, which diffuse across a 100–200 Å synaptic cleft and react with receptors on the postsynaptic motoneuronal membrane. By unknown processes (probably involving macromolecular conformational changes (39)) this reaction leads to an increase in the local permeability of the membrane to certain ions, Na^+ and K^+ at excitatory synapses (13) and Cl^- at inhibitory synapses (35–37). The identities of the excitatory and inhibitory transmitters are not yet known (see ref 29 for likely candidates).

At several central synapses the flow of ionic current across the activated postsynaptic membrane is described approximately by the empirical equation $I_s = g_s(t)$ $(V_{rev} - V(t))$, where I_s is the net current flowing into the neuron at the particular synapse, $g_s(t)$ is the synaptic conductance change associated with the change in membrane ionic permeability, $V(t)$ is the electrical potential difference across the cell membrane at time t, and V_{rev} is the reversal potential of the particular synaptic current (13). The quantity $(V_{rev} - V(t))$ is called the synaptic driving potential. Synaptic current flow is thus proportional to the synaptic conductance change and to the synaptic driving potential, and acts to bring the postsynaptic membrane potential toward V_{rev}.

The reversal potential for a particular synaptic event is determined by the relative ionic permeabilities of the activated synaptic membrane and by the concentrations (both inside and outside the neuron) of the ions involved in the permeability changes, as indicated by the Goldman equation (8). For excitatory synaptic events V_{rev} is close to 0 mV. An excitatory synaptic event impinging on a resting motoneuron ($V_{rest} = -70$ mV) will depolarize the cell, driving its transmembrane potential towards 0 mV. For inhibitory synapses V_{rev} is about -80 mV, so an inhibitory synaptic event will hyperpolarize the motoneuron.

Compared to the input conductance of the motoneuronal soma (about 10^{-6} mho) the conductance change caused by a single (quantal) excitatory or inhibitory transmitter packet is very small. Thus these miniature synaptic events produce only a small voltage change at the cell body of the neuron (about 100 μV for an excitatory quantum from Ia afferent synapses (32, 33)). Many excitatory synaptic events must sum together to depolarize the soma to the threshold voltage (about -60 mV, 10 mV positive to the resting potential), at which the motoneuron fires its output message, the action potential.

Action potential generation

The action potential is a rapid transient depolarization, produced by sequential, voltage-dependent increases in membrane permeability to Na^+ and K^+. These voltage-dependent permeability changes are distinct from the transmitter-sensitive permeability to Na^+ and K^+ that mediates excitatory synaptic transmission. In motoneurons the action potential is initiated at the region of lowest threshold, probably the initial portion of the axon (13). From this region the action potential propagates both "forwards" down the axon toward the neuromuscular synapse and "backwards" into the cell body and dendrites of the motoneuron. The extent to which the action potential invades the motoneuron dendrites is still controversial.

The detailed mechanism of action potential generation in central neurons is complicated, and at present is not very well understood, even for the extensively studied motoneuron. For this discussion we will instead use the empirical observation that, once the voltage at the soma exceeds threshold, the firing rate (action potential output rate) of the motoneuron is roughly proportional to the magnitude of the net depolarizing current reaching the soma from all the synaptic events on the neuron (16, 17). Thus, the motoneuron may be thought of as an integrator of many synaptic inputs, with a time constant of about 6 msec. The above approximation is reasonably accurate, but there are deviations due to such phenomena as accommodation and adaptation, and to nonlinear interactions between neighboring synaptic conductance changes.

PASSIVE MEMBRANE PROPERTIES

The synaptic interactions that determine the net current reaching the soma are complex because motoneurons have an intricate geometry and are not isopotential. Of the 20,000–50,000 synaptic terminals ending on the alpha motoneuronal membrane, over 80% occur on the dendritic processes of the neuron (calculated by multiplying synaptic density measurements (9) by surface area measurements (3)). Dendrites have a lower input conductance than the soma, and a quantal excitatory synaptic conductance change on a dendrite produces a large local depolarization, up to 20 mV in fine distal dendrites (4). Such a large depolarization would elicit an action potential if it occurred on the soma, but the dendrites of normal motoneurons are relatively inexcitable (i.e., they have a much higher threshold than the soma). Thus the local dendritic depolarization usually spreads passively along the dendritic branches toward the soma. This passive spread of potential is called electrotonic propagation (18), and is described using concepts and equations originally applied to current flow in undersea cables (8). Thus the dendrites are often treated mathematically as core conductors, and the passive resistive and capacitive properties of the motoneuronal membrane and cytoplasm are called the cable properties of the motoneuron.

Only a fraction of the synaptic current (or charge) injected at a dendritic synapse will reach the soma and thereby have an effect on motoneuronal output. The remainder of the current will leak out through the intervening dendritic membranes. The cable properties and geometry of the motoneuron determine the "effective" fraction of the synaptic current from a particular dendritic site. The lower the cytoplasmic resistivity (R_a), the higher the specific membrane resistance (R_m), and the more proximal the synaptic site, the more effective a particular synapse will be in influencing motoneuronal output. Thus to calculate the relative effectiveness of dendritic synapses, it is necessary to measure R_a, R_m, and the dendritic geometry of the neuron. Unless otherwise noted, computation methods and results are described in detail in refs 3 and 4.

Estimates of cytoplasmic resistivity, R_a

The calculations of R_m and synaptic efficacy to be described here require a value for the longitudinal resistivity of dendritic cytoplasm. To date this property has not been measured accurately. The minimal reasonable value for R_a is 50 Ωcm, the resistivity of mammalian balanced salt solution (Tyrode's solution) at 37 C. Crude estimates of the cytoplasmic resistivity of the motoneuron soma averaged 70 Ωcm (3), but somatic cytoplasm may have an effective resistivity greater than that of dendritic cytoplasm because of the convoluted membranes of the somatic endoplasmic reticulum. Most of the organelles in the dendrites (microtubules, neurofilaments) run longitudinally and are not densely packed; thus they would not be expected to impede longitudinal current flow. The studies reported here assumed R_a values between 50–100 Ωcm; within this range errors in R_a would alter calculated values of R_m by no more than 30%.

Estimates of specific membrane resistance and neuronal geometry

Membrane resistance is calculated using the geometry of the neuron and the measured input resistance of at least one site of the neuron (usually the soma). The input resistance at the soma is evaluated as the slope of the relation between the steady-state change in somatic voltage produced by injecting a subthreshold current step into the soma, and the magnitude of that current. Assuming uniform passive membrane properties throughout the neuron (see below), one then calculates the predicted input resistance of the soma for various assumed values of R_m. The R_m value which gives a predicted input resistance equal to that actually measured is taken as the uniform R_m of the neuron (41). The cat motoneuron R_m values calculated in early studies varied over a wide range. One early estimate of R_m, 600 Ωcm² (10), was so low as to suggest that practically all the current from synaptic events originating on the distant dendrites would leak out through the dendritic membranes, and never reach the soma. Thus distant dendritic synapses were thought to have almost no influence on motoneuronal function. Later calculations by Rall (41) suggested higher R_m values between 2,000–8,000 Ωcm², and predicted that dendritic synapses could produce significant soma depolarization if many were active together.

The variation in reported R_m values was due in large part to the methods used to estimate dendritic geometry. Studies reporting low R_m values estimated neuronal geometry from Golgi-stained spinal cord neurons. The alcohol dehydration used with the Golgi technique shrinks neurons (up to 50% by volume), and neuronal processes that leave the histological section are lost. Both these artifacts reduce the calculated area of the dendritic membrane, and reduce the calculated value of R_m. Another source of error resulted from making geometrical and electrical measurements on different populations of motoneurons.

Recent technical advances have made it possible to determine the geometry of the same motoneurons studied electrophysiologically. Lux et al. (38) injected motoneurons with radioactive glycine and then reconstructed the neurons from serial autoradiographs. Their geometrical measurements suggested that the dendritic tree could be approximated by finite, equivalent, core-conducting cylinders. Using this equivalent cylinder approximation, they calculated an average R_m value of 2,700 Ωcm^2. W. Crill and I injected motoneurons with a fluorescent dye, Procion yellow, and used histological procedures that minimized cell shrinkage. Our reconstructions of the dye-filled motoneurons (Fig. 1) indicated that dendrites end at different electrical distances from the soma, and that individual dendrites decrease in diameter over much of their length. These measurements did not support an equivalent cylinder approximation of the dendritic tree, so we instead approximated the dendrites in a more detailed way, as a series of 300–700 short, interconnected membrane cylinders, as proposed by Rall (41) (see Fig. 2). Computer-assisted calculations applied to this detailed geometrical approximation yielded R_m values ranging between 1,200 and 3,500 Ωcm^2, averaging about 2,000 Ωcm^2.

The calculations of R_m cited above assume that R_m is uniform over both somatic and dendritic membranes. Since the value of dendritic R_m is critical in calculating the effectiveness of dendritic synapses, it is important to determine how much dendritic R_m values might differ from the uniform R_m value of 2,000 Ωcm^2. Calculations assuming different R_m

values for somatic and dendritic membranes indicate that the uniform R_m value is close to the lower bound for the average, *resting* value of dendritic R_m. Dendrites constitute more than 80% of the motoneuronal membrane (excluding the axon), and if dendritic R_m were much less than 2,000 Ωcm^2, the neuron input resistance could not be as high as the values measured electrophysiologically, even if somatic R_m were infinite.

Other evidence supporting a high R_m value comes from studies that analyzed the transient voltage response to a current step injected into the motoneuron soma. The voltage response of a finite core conductor model of the dendritic tree may be expressed as a series of exponential terms. The ratio of the time constants of the two slowest terms in this exponential series is related to the electrical length of the dendritic cable (equation 3, ref 43). Using Rall's equations, Lux et al. (38) and Burke and ten Bruggencate (7) calculated that the average electrical length of motoneuron dendrites ranges from 1.1 to 2.2, averaging about 1.5 space constants. These values correspond quite well to the electrical lengths of dendrites calculated from motoneuronal geometry using the recently determined high R_m values (3, 38). If the true value of dendritic R_m were significantly lower than 2,000 Ωcm^2, the average dendritic electrical length calculated from motoneuronal geometry would have been considerably shorter than that calculated from the transient response.

It is more difficult to place an upper bound on dendritic R_m. One suggestion that dendritic R_m may be higher than 2,000 Ωcm^2 comes from calculating the specific membrane capacitance, C_m. If R_m is uniform over all parts of the resting neuronal membrane, then the longest time constant τ_0 of the passive response of the neuron to a current step should be equal to the product of R_m and C_m (18). Using this relationship, measured values of τ_0 and calculated values of R_m give C_m values averaging 2.5 $\mu F/cm^2$. While this C_m value is considerably lower than earlier estimates, it is still higher than the value of 1 $\mu F/cm^2$ found in most biological membranes. Some highly convoluted membranes have higher apparent C_m values (15, 40), but electron micrographs show that the motoneuron membrane is not greatly convoluted. Thus the high calculated C_m value hints that R_m may not be uniform over the resting neuron, and in fact, within experimental error, the transient voltage response of the soma to local injection of a current step can be

fit as well by assuming a very high dendritic R_m (8,000 Ωcm^2), a low soma R_m (250 Ωcm^2) and a "standard" C_m (1 $\mu F/cm^2$) as by assuming a uniform R_m of 2,000 Ωcm^2 and a higher C_m (2.5 $\mu F/cm^2$). Consequently, it is possible that the resting value of dendritic R_m may be considerably higher than 2,000 Ωcm^2, and even the soma may have a higher R_m when it is not damaged by microelectrode penetration.

CHARGE INJECTION AT DENDRITIC SYNAPSES

Quantal components of synaptic events

In order to evaluate the role of dendritic synapses in motoneuronal function it is also important to know the

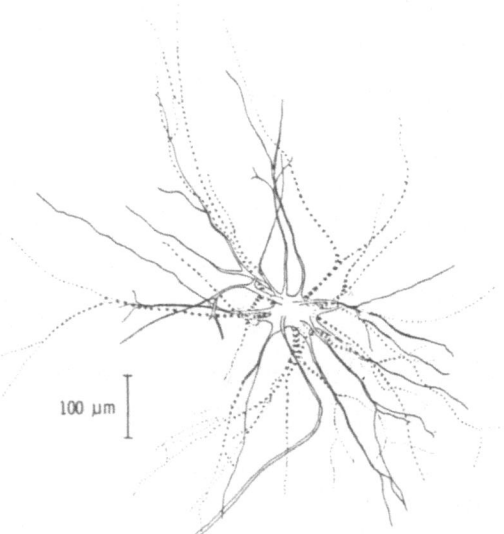

100 μm

Figure 1. Reconstruction of motoneuron injected with Procion yellow dye. This neuron has 19 primary dendrites and a total membrane surface area of 180,000 μm^2, and was reconstructed from 12 serial cross sections of the spinal cord. Dendrites indicated by dashed lines project behind the plane of the soma. The input resistance measured at the soma was 1.5 MΩ. Calibration, 100 microns. Reprinted from ref. 3.

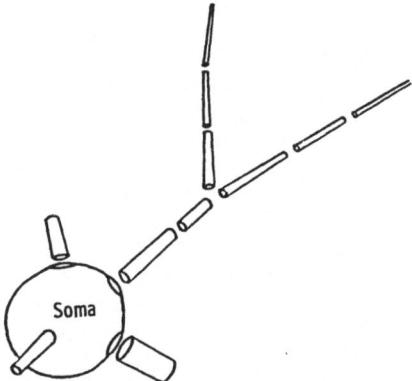

Figure 2. Schematic diagram of the soma and a dendritic branch, illustrating how dendrites reconstructed as in Fig. 1 were approximated as a series of interconnected cylinders, to facilitate calculation of membrane properties and synaptic effectiveness. Calculation procedures are described in (3).

magnitude and time course of the synaptic conductance change. Evidence reviewed briefly below suggests that the nerve terminals that synapse with motoneurons release their chemical transmitters in multimolecular packets, called quanta. Thus our calculations require estimates of both the size of a single (quantal) conductance change and the number of quanta released at the synaptic site.

The firmest support for the quantal theory of chemical synaptic transmission comes from studies of acetylcholine release from motoneuron terminals at the nerve-muscle junction (reviewed in 19, 24). Morphological and physiological evidence suggests that the synapses on motoneurons also release their transmitters in quantal packets. Electron micrographs of nerve terminals synapsing on motoneurons contain many small vesicles, each thought to store one quantal unit of transmitter. Electrophysiological recordings from motoneurons in frogs and cats show

small spontaneous synaptic potentials that persist when nerve impulse activity is blocked (analogous to spontaneous quantal end-plate potentials at the nerve-muscle junction (5, 6, 25)), and the EPSPs (excitatory postsynaptic potentials) evoked in motoneurons by stimulating single Ia afferent fibers fluctuate in size in a manner compatible with the quantal hypothesis (30). The widespread association of synaptic vesicles with axon endings suggests that transmitter release at other central synapses also occurs in quantal packets, but supporting physiological evidence of quantal transmission is still scanty.

Data of Kuno and Miyahara (32, 33) suggest that the size of a quantal EPSP originating on the motoneuron soma is about 100 μV in a soma with an input resistance of 1 MΩ. The conductance change required to produce this quantal EPSP has a peak amplitude of $90 - 180 \times 10^{-10}$ mho, a duration of about 0.3 msec, and a time integral of 40×10^{-10} mho-msec (4).

The calculations presented below assume a quantal conductance change of similar magnitude and time course for dendritic synapses. Most of the Ia afferent terminals that synapse on motoneurons release no more than 1–5 quanta per impulse (30, 33). The average quantal output at most other motoneuronal synapses is unknown, but is assumed to be similarly low.

Reduced synaptic charge injection at dendritic synapses

The 10^{-8} mho conductance change underlying a quantal EPSP is less than half the size of a quantal conductance change at the frog neuromuscular junction (11, 14), but it is a large conductance change compared to the low input conductance of about 5×10^{-9} mho calculated for fine distal dendrites (in this and following sections, "distal" refers to dendritic branches more than 1 space constant (400–1,000 μm) from the soma). Calculations using the core conductor motoneuron model predict local quantal potential changes as large as 20 mV at synaptic sites on distal dendrites (4). These synaptic potentials are great enough to reduce the synaptic driving potential $(V_{rev} - V(t))$ significantly, even during the brief synaptic conductance change. This reduction in driving potential reduces the synaptic current $(I_s = g_s (V_{rev} - V(t))$, so less total charge is injected into the neuron by the conductance change. Calculations indicate that the amount of charge injected into the subsynaptic region by a quantal conductance change occurring on a distal dendrite is only 80% of that injected by a similar conductance change occurring on the soma.

When several quantal conductance changes occur simultaneously or within a membrane time constant (approximately 6 msec) of each other at the same or closely neighboring sites on a distal dendritic branch, the average synaptic driving potential is further reduced, and the amount of charge injected per unit conductance change decreases correspondingly. For example, four quantal packets of transmitter released simultaneously at the same distal dendritic site will inject only 50% of the synaptic charge associated with a similar conductance change occurring on the soma, with its higher input conductance. Thus, nearly synchronous neighboring synaptic conductance changes on distal dendrites interact in a nonlinear way, such that the total amount of charge injected by the synchronous conductance changes is less than the total amount of charge injected when the same conductance changes occur asynchronously. This less-than-linear

summation of quantal synaptic events on motoneuron dendrites was first observed by Kuno and Miyahara (32). The degree of nonlinear summation they observed indicated that some of the quantal EPSPs were as large as 15 mV at their dendritic site of origin. Two synapses located on different dendrites are electrically isolated from each other, and hence will not interact in this nonlinear way. To summarize, the phenomenon of less-than-linear summation operates at low-input-conductance dendritic sites to reduce the amount of charge injected both by a single quantal conductance change, and by multiple conductance changes occurring together in space and time.

CURRENT FLOW FROM DENDRITE TO SOMA

In order to influence motoneuronal output, synaptic currents must reach the soma. Some of the charge injected at dendritic synaptic sites leaks out through the dendritic membranes before reaching the soma. The fraction of the injected synaptic charge that reaches the soma from a particular dendritic site can be calculated by iteratively applying the cable equation to segments of a detailed core conductor model of the neuron (as in Fig. 2, see ref 41). For subthreshold voltage changes, this fraction is independent of the time course of charge injection, because the motoneuron may be approximated as a linear system between the resting potential and threshold (3, 4, 22, 23, 41). The results of applying this calculation procedure to a motoneuron dendrite are illustrated by the solid circles in Fig. 3. These calculations used the resting R_m value of 2,000 Ωcm², and it is evident that at this R_m, most of the current injected at dendritic synapses does reach the soma. Current injected into even the most distal

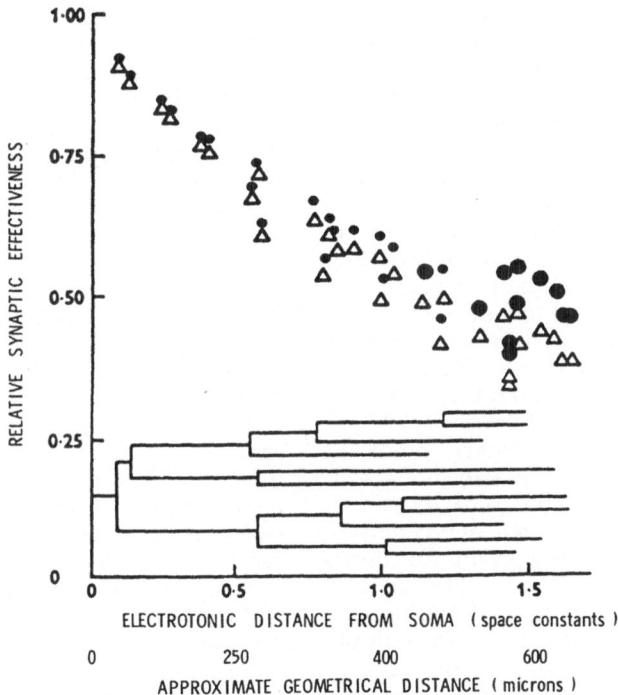

Figure 3. Relative synaptic effectiveness as a function of distance from the soma, calculated for a reconstructed dendrite (illustrated schematically by the inset line drawing) in a resting motoneuron ($R_m = 2,000 \ \Omega cm^2$). Filled circles indicate the fraction of injected charge reaching the soma from each site; larger circles represent values calculated for terminal dendritic branches. Open triangles plot the relative effectiveness of quantal excitatory synaptic conductance changes in contributing to the net depolarization of the soma. For a given dendritic site the difference between the circle and triangle reflects reduced charge injection due to the reduction in synaptic driving potential that occurs during the synaptic conductance change (nonlinear summation). See text and ref 4 for further details.

dendritic synapses will be at least 28 to 50% as effective in depolarizing the soma as current injected directly into the soma.

RELATIVE EFFECTIVENESS OF DENDRITIC SYNAPSES

The contribution of dendritic synapses to motoneuronal function is limited by the factors discussed in the two preceding sections, *1*) the reduction in synaptic current injection due to decreased synaptic driving potential (nonlinear summation), and *2*) the leakage of current through the membranes between the synaptic site and the soma. Both these factors become more important with increasing distance from the soma. By combining these two factors, we arrive at a measure of the relative effectiveness of a dendritic synaptic site, defined as the ratio of the net soma depolarization in time, $\int_0^\infty V_s \ (t) \ dt$, produced by a given synaptic conductance change at the dendritic site, to the soma depolarization that would have been produced had that conductance change occurred directly on the soma. The open triangles in Fig. 3 show the relative effectiveness of an excitatory quantal conductance change as a function of location along a dendrite. Even the most distal sites on this dendrite are at least 30% as effective as somatic synapses. In a sample of 10 dye-injected motoneurons, the most distant synapse was still at least 20% as effective as a somatic synapse, and quantal conductance changes occurring on 76% of the dendritic tree were predicted to be at least 50% as effective as synapses directly on the soma. These calculations suggest that dendritic synapses have an important functional role in motoneurons. The relative effectiveness of distal dendritic sites will decrease, of course, for multiple synchronous quantal conductance changes (because of increased nonlinear summation), and will also decrease as neighboring synaptic events reduce R_m and the synaptic driving potential (see below).

Data from a recent study (21) suggest that some dendritic synapses may be even *more* effective than the above calculations predict: Ia afferent fibers synapsing on the more distal portions of motoneuron dendrites seem to inject *more* charge into the soma than similar afferents synapsing more proximally. The reasons for this surprising finding are still unknown. Possible explanations include a greater quantal conductance change at distal synapses, a larger quantal output per impulse, and/or a larger number of terminal boutons per afferent fiber at distal compared to proximal synapses, or even partial active responses in distal dendrites.

The EPSP produced at the soma by dendritic synapses does have a slower rising phase than an EPSP originating on the soma, due to the cable properties of the dendrites (42). However, the initiation of action potentials in cat lumbosacral motoneurons does not seem to be differentially sensitive to the faster components of the EPSP (4). Thus the net somatic depolarization produced by an EPSP is a good measure of its contribution to the firing rate of the motoneuron (4, 17, 21).

In other neurons action potential initiation may be more sensitive to the faster components of synaptic potentials. In order to evaluate the effectiveness of dendritic synapses in these neurons, it would be necessary to calculate the shape or the frequency power spectrum of the potential change produced at the soma (or better, at the action potential initiating region) by synaptic events at different dendritic locations. These calculations could be done either by using multicompartment neuron models (42) or by applying linear systems methods to calculate the attenuation associated with each frequency component of the dendritic EPSP (2, 44, 45).

REDUCTION OF EFFECTIVE MEMBRANE RESISTANCE BY SYNAPTIC CONDUCTANCE CHANGES

Are the synaptic conductance changes that occur during normal motoneuronal functioning ever great enough to reduce the effective value of the specific membrane resistance? It is difficult to answer this question at present because the many different synaptic inputs to spinal motoneurons show a wide range of activity in different functional circumstances. However, it is possible to estimate the effective values of the specific membrane resistance during some situations which are expected to occur during normal functioning.

The total specific membrane conductance, G_T (the conductance per unit area of membrane), is equal to the sum of the resting membrane conductance, G_m ($G_m = 1/R_m$), and any synaptic conductance changes, G_s:

$$G_T = G_m + G_s$$

The synaptic conductance per unit area of membrane, G_s, is determined by the rate of impulse activity in presynaptic terminals, the amount of transmitter released by each impulse, the magnitude and time course of the postsynaptic conductance change produced by each quantal packet of transmitter, and the number of active synaptic boutons per unit area of motoneuron membrane. Estimation of G_s is further complicated by the heterogeneity of synapses on motoneurons. The relatively well studied dorsal root Ia afferent input accounts for fewer than 5% of the synaptic boutons on motoneurons (9). The quantal conductance change associated with at least some inhibitory synapses on motoneurons is three times greater than the quantal conductance change associated with the excitatory Ia input discussed above (34). In addition, it seems unlikely that all presynaptic fibers release the same average number of quanta per impulse, and fibers from different sources may synapse preferentially on restricted areas of the motoneuron.

Keeping in mind these various complexities, G_s can be estimated using three different methods: 1) by calculating G_s for various rates of afferent input assuming that all synapses on motoneurons produce, on the average, the same magnitude quantal conductance change as Ia afferent synapses, 2) by directly measuring the input conductance of the neuron during a given synaptic bombardment and then calculating the effective membrane conductance, and 3) by calculating the minimal excitatory synaptic conductance change required to depolarize the soma enough to produce action potentials at the maximum rate observed physiologically. The results of applying each of these methods are described below.

1) Estimating G_s by extrapolating data from Ia synapses

If, on the average, all presynaptic terminals produce the same quantal conductance change as the Ia terminals, G_s will be a function of the average quantal release rate per synaptic bouton, f:

$$G_s = A \cdot f,$$

where the proportionality constant A is the product of the density of synaptic boutons (20/100 μm^2 or $2 \times 10^5/$cm^2, ref 9), the unit used for membrane area (cm^2) and the time integral of the quantal synaptic conductance change (about 40×10^{-13} mho-sec).

Figure 4 plots the effective value of membrane resistance, R_T ($R_T = (G_T)^{-1} = (G_m + G_s)^{-1}$), as a function of G_s (upper abscissa) and of the transmitter release rate (in quanta per second per synaptic bouton, lower abscissa), calculated using this assumption. The solid curve was calculated using a resting R_m value of 2,000 Ωcm^2 (a lower bound on R_m, see above), and the dotted curve was calculated assuming a resting R_m value of infinity (an upper bound). Calculated R_T values decrease rapidly with increasing rates of synaptic activity. If the greater quantal conductance change associated with some inhibitory synapses (34) were used instead, the quantal release rates required to produce a given reduction of R_T would be only one third of the rates given in Fig. 4.

It is of course difficult to determine the number of quanta actually released from each synaptic bouton per presynaptic action potential. The data of Kuno (30) suggest that an action potential in a single Ia fiber releases an average of about one quantum of transmitter, but the number of synaptic boutons ending on a single motoneuron from a single afferent fiber is not known, and probably varies considerably for different presynaptic axons. However, afferent axons and interneurons can generate action potentials at rates exceeding 200/sec during normal functioning, so even if release rates averaged only

0.1 quantum per action potential per bouton, such a uniform intense barrage could reduce R_T substantially (to about 500 Ωcm² in Fig. 4). These calculations are of course speculative, but Fig. 4 does indicate that reasonable rates of synaptic activity can significantly reduce the effective membrane resistance of motoneurons. The dotted curve, calculated assuming a resting R_m of infinity, suggests that even the so-called resting value of R_m may have been strongly influenced by the low rates of synaptic activity that survived the bar-

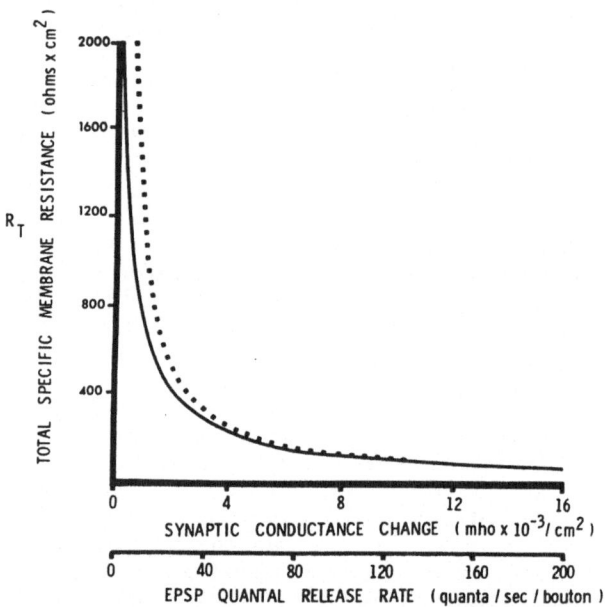

Figure 4. Total effective specific membrane resistance, R_T, as a function of the synaptic conductance change per unit membrane area. The solid curve was calculated assuming a resting ($G_s = 0$) specific membrane resistance of 2,000 Ωcm² (a lower bound, see text); the dotted curve, assuming an infinite resting specific membrane resistance. The lower abscissa gives the average excitatory quantal release rates required to produce a given synaptic conductance change. The EPSP quantal release rates were calculated assuming that all synaptic boutons on the motoneuron surface produce quantal conductance changes similar to those produced on motoneurons by Ia afferent fibers. The similarity of the solid and dotted curves suggests that R_T is determined primarily by synaptic conductance changes when the average background synaptic activity exceeds a few quanta per second per bouton.

biturate anesthesia used during the determination of R_m (see ref 46). In the complete absence of any synaptic activity R_m could well be considerably higher than 2,000 Ωcm^2.

2) Measuring G_T during synaptic activity

Simultaneous stimulation of both the descending fiber tracts of the spinal cord and the dorsal roots produces a great reduction in the motoneuron input resistance measured at the soma. The intensity of the stimuli applied to the various tracts and roots can be balanced to give a mixture of excitation and inhibition that changes the membrane voltage at the soma by only a few millivolts, while drastically reducing the measured input resistance. The effective value of specific membrane resistance, R_T ($R_T = (G_T)^{-1}$), corresponding to this reduced input resistance can be calculated using the methods employed previously to calculate R_m in the resting neuron. Such calculations show that R_T decreases markedly with increasing frequency of stimulation, such that when the descending tracts and dorsal roots are stimulated at 200/sec, R_T drops to just 100 Ωcm^2, only 5% of its resting value. This calculation assumes a spatially uniform R_T, a condition not likely to be met during this synaptic bombardment, but the membrane resistance must drop this low in at least some regions of the motoneuron membrane.

This massive, widespread stimulation activates many interneurons and inhibitory synapses, so the frequency of presynaptic action potentials and the average quantal release rate during such stimulation are not known. However, even this rather crude experiment shows that synaptic conductance changes can greatly reduce R_T. Further experiments of

this kind, but using morphologically characterized, discrete inputs that do not activate interneuron pools, might make it possible to relate the measured synaptic conductance changes in motoneurons to the rate at which action potentials invade the presynaptic boutons.

3) Excitatory conductance changes required to produce maximal rates of action potential output in motoneurons

During normal functioning cat motoneurons generate action potentials at rates up to 100/sec. This maximal rate can be achieved by injecting about 50 nA of depolarizing current into the soma (for motoneurons with an input resistance of about 1 MΩ, (26–28)). If no inhibitory synapses are active, the minimal rate of excitatory quantal conductance changes required to maintain a 50 nA depolarizing current can be calculated as described below.

When the motoneuron is firing action potentials at its maximal rate, whether due to synaptically or artificially injected current, the average depolarization at the soma will be at least 20 mV. Thus the average driving potential for an EPSP originating at the soma will be reduced to 50 mV or less from the resting value of 70 mV. The net excitatory synaptic conductance change, g_s, required to generate the 50 nA of current needed to maintain maximal firing rates can be calculated by dividing this current by the reduced synaptic driving potential:

$$g_s = \frac{I_s}{V_{rev} - V(t)} = \frac{50 \times 10^{-9} \text{ amp}}{50 \times 10^{-3} \text{ volt}} = 10^{-6} \text{ mho}$$

This simplified calculation underestimates G_s, by assuming that all EPSPs originate on the soma. If instead the EPSP activity is distributed uniformly over somatic and dendritic membranes, the required g_s will be no more than twice that calculated above, since the average dendritic synapse will still be at least 50% as effective as somatic synapses under these conditions (see below). Dividing this total synaptic conductance change by the total membrane area of a motoneuron (about 2×10^{-3} cm²), yields a normalized conductance change, G_s, of $0.5 - 1 \times 10^{-3}$ mho/cm². By reference to Fig. 4, this normalized conductance change would reduce R_T from 2,000 Ωcm² to 1,100–700 Ωcm², and could be produced by the release of only 6–12 quanta/sec per bouton (assuming that all boutons are excitatory, with an average quantal conductance change similar to that described above for Ia afferent fibers). If only the Ia afferents were active, they would have to release quanta at 20 times this rate, 120–240/sec, to produce maximal rates of action potential generation, since they constitute only about 5% of the presynaptic terminals on cat motoneurons.

In the presence of simultaneous inhibitory conductance changes, the excitatory conductance changes required to produce maximal firing would be much greater. Assuming again that the average depolarization of a rapidly firing motoneuron is at least 20 mV, the magnitude of the driving potential for IPSPs will be 30 mV ($V_{rev} = -80$ mV), 60% of the magnitude of the EPSP driving potential. Thus, an additional excitatory conductance change equal to at least 60% of the added inhibitory conductance change would be required to maintain the same rate of action potential generation. In fact, the compensating excitatory conductance change would have to be even greater, because the added inhibitory conductance changes also reduce the effective membrane resistance R_T, which reduces the relative effectiveness of dendritic synapses (see below).

In summary, the results of the calculations presented above indicate that synaptic conductance changes can greatly reduce the effective specific membrane resistance. The purely excitatory conductance change required to drive the motoneuron to its maximal rate of action potential generation would reduce R_T from 2,000 to about 700–1,100 Ωcm². If inhibitory synapses are active simultaneously, the reduction of R_T corresponding to a particular motoneuronal firing rate will be even greater.

REDUCTION OF SYNAPTIC EFFECTIVENESS BY BACKGROUND SYNAPTIC ACTIVITY

The relative effectiveness of a quantal synaptic conductance change on a dendrite is strongly influenced by the other synaptic activity occurring on that dendrite at the same time. Background synaptic activity will change the effectiveness of a test quantal conductance change both by altering its driving potential (due to the large local changes in membrane potential that accompany synaptic activity on the dendrites), and by reducing the effective value of the dendritic membrane resistance, hence increasing the fraction of synaptic charge lost through the dendritic membranes. Since many synapses are active during normal functioning, the quantitative aspects of these synaptic interactions are important.

The actual background synaptic activity on a given dendrite is composed of quantal synaptic conductance changes, each having a discrete time course. Thus the background synaptic conductance will fluctuate in time, even when the average rate of afferent input to the motoneuron is approximately constant. As a first approximation this fluctuating background synaptic activity can be characterized as a steady conductance change equal to the average value of the fluctuating synaptic conductance. Preliminary computations by M. Chujo and myself indicate that the average relative effectiveness of a test quantal EPSP is slightly greater in the presence of a fluctuating background than during this steady average background conductance change. Thus, calculations assuming a steady average background conductance change should set a lower limit on the average relative effectiveness of a quantal conductance change in the presence of fluctuating background synaptic activity.

Figure 5 illustrates the effect of the steady background conductance change on the synaptic effectiveness of EPSPs originating on the soma, and on dendritic branches 100, 255, and 650 microns from the soma of a reconstructed motoneuron. The calculations approximated the dendrites as a series of interconnected, core conducting cylinders, as described above and illustrated in Fig. 2. The large depolarizations accompanying intense synaptic activity and the resulting action potentials will both activate the voltage-dependent membrane conductance to K^+. The calculations assumed that this K^+ conductance change is restricted to the soma membrane, and used the current-voltage relationships measured directly from motoneurons by Araki and Terzuolo (1) to specify the soma voltage at any given net synaptic current. If this voltage-dependent, K^+ conductance system actually extends out into dendritic membranes, these calculations will overestimate

the relative effectiveness of dendritic synapses in cases where dendritic depolarizations are present.

The dotted curves in Fig. 5 plot the relative synaptic effectiveness calculated assuming that the background synaptic activity produces no change in the membrane potential of the motoneuron (as might occur with a mixed excitatory-inhibitory input). The dotted curves thus show the effect of reduced R_T, in the absence of any reduction in synaptic driving potential. The solid curves were calculated assuming that the background synaptic conductance change is purely excitatory with a reversal potential of 0 mV. The difference between the solid and dotted curves is thus due entirely to reduction of the synaptic driving potential by the background excitatory activity. The abscissa of Fig. 5 includes both the conductance change per unit area of membrane and the excitatory and inhibitory quantal release rates required to produce that conductance change.

Simplified calculations described in an earlier section indicated that the minimal purely excitatory conductance change required to drive a motoneuron to its maximum firing rate is $0.5-1 \times 10^{-3}$ mho/cm². From Fig. 5 (solid curves) it is evident that this conductance change does not greatly reduce the relative effectiveness of the two more proximal dendritic sites, but does reduce the relative effectiveness of the most distal site to $0.15-0.07$, only 30 to 14% of its resting value of 0.5. As mentioned above, the relative effectiveness would be slightly greater for a fluctuating background conductance change. These calculations show that a purely excitatory input, either uniformly distributed or concentrated at the soma, could drive a motoneuron to its maximal firing rate with

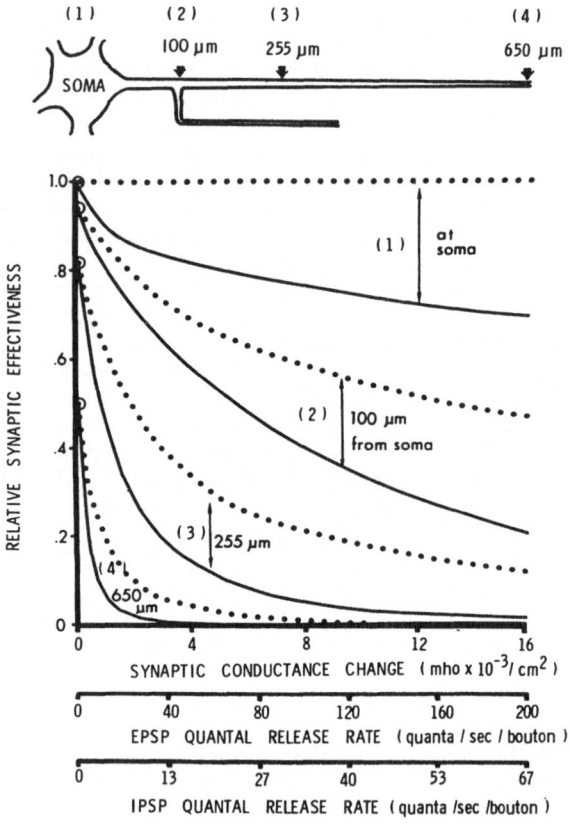

Figure 5. Paired curves give the relative synaptic effectiveness of a test excitatory quantal conductance change at four synaptic sites as a function of the background synaptic conductance change. The synaptic sites, labeled in the schematic diagram above, are *1)* on the soma and *2)* 100 μm, *3)* 255 μm and *4)* 650 μm from the soma along a typical reconstructed dendrite. Calculation of the dotted curves assumed that the net reversal potential of the background synaptic conductance change was −70 mV, i.e., that the background synaptic activity produced no change in the transmembrane potential of the neuron. Calculation of the solid curves assumed a background reversal potential of 0 mV, i.e., that the background synaptic activity was purely excitatory. For both sets of curves the reversal potential of the test quantal event was 0 mV, and the background conductance change was assumed to be distributed uniformly over the neuronal membrane. The lower two abscissas give the excitatory and inhibitory transmitter release rates required to produce the synaptic conductance changes given by the upper abscissa. The EPSP quantal release rate assumes that all synaptic boutons are excitatory, with an average quantal conductance change similar to that described for Ia afferent fibers. The background EPSP rates apply only to the solid lines, which include a reduction in the EPSP driving potential. The IPSP quantal release rate assumes that all but the test bouton are inhibitory, with a quantal conductance change three times larger than that for EPSPs (34). The background IPSP rates apply most closely to the dotted curves, calculated assuming no reduction in the EPSP driving potential. Note that an increase in the background synaptic conductance change reduces the relative effectiveness of distal synapses much more severely than that of proximal synapses.

little change in the relative effectiveness of all but the most distal dendritic sites.

'In contrast, the conductance changes associated with an inhibitory input or a mixed excitatory-inhibitory input can be sufficient to reduce drastically the relative effectiveness of dendritic synapses, even at slow motoneuronal firing rates. Intense stimulation of dorsal roots and descending afferent tracts can reduce R_T to 100 Ωcm^2 (see above), which corresponds to an average synaptic conductance change, G_s, exceeding 10^{-2} mho/cm^2, equivalent to the release of 40 inhibitory quanta/sec per bouton. Figure 5 (dotted curve) shows that such conductance changes would render any excitatory synaptic activity on the distal dendrites completely ineffective in depolarizing the soma, and the relative effectiveness of the intermediate dendritic site (255 μm from the soma) is reduced to about 0.18, 22% of its value in the resting neuron. Only synapses on the soma and the proximal 100 μm of the dendrites would retain more than 50% of their resting effectiveness. Another example of the powerful shunting effect of inhibitory synapses is Diamond's (12) observation that application of GABA (an inhibitory transmitter) to one dendrite of a goldfish Mauthner cell produced a conductance increase large enough to render all other synaptic inputs on the distal part of that dendrite ineffective in depolarizing the soma. Note that any uniform synaptic conductance increase differentially decreases the relative effectiveness of the more distal synapses. The more proximal the dendritic site, the greater the synaptic conductance change required to reduce its relative effectiveness significantly. Excitatory synaptic sites on the soma or proximal dendrites effectively deliver depolarizing currents

to the soma even in the presence of intense background conductance changes.

It seems unlikely that every afferent input distributes its presynaptic terminals uniformly over dendritic and somatic membranes. Nonuniform terminal distributions would have an important effect on synaptic integration in motoneurons. For example, inhibitory synapses on a distal dendritic branch would differentially reduce the effectiveness of excitatory synapses on that branch, with little effect on synapses on different dendrites. In contrast, inhibitory synapses localized on the soma would uniformly reduce the effectiveness of all excitatory synapses on the neuron.

CONCLUSION

The dendritic tree vastly increases the membrane area available for synaptic input into motoneurons. Recent measurements of the electrical and geometrical properties of cat motoneurons indicate that the average specific resistance of the dendritic membranes is at least 2,000 Ωcm^2 in the resting neuron. Calculations using this value predict that quantal excitatory conductance changes occurring on over 75% of the dendritic tree will be at least 50% as effective as somatic synapses in depolarizing the soma, and that even the most distal dendritic synapses will be at least 20% as effective as somatic synapses. Thus dendritic synapses have a critical role in motoneuronal integration. However, the rates of synaptic activity present during certain periods of normal functioning may greatly reduce the relative effectiveness of excitatory dendritic synapses, both by reducing the synaptic driving potential and by increasing the leakage of synaptic current through dendritic membranes.

I thank Dr. E. Barrett for reading and editing the manuscript.

REFERENCES

1. ARAKI, T., AND C. A. TERZUOLO. Membrane currents in spinal motoneurons associated with the action potential and synaptic activity. *J. Neurophysiol.* 25: 772, 1962.
2. BARRETT, J. N. Determination of neuronal membrane properties using intracellular staining techniques. In: *Intracellular Staining Techniques in Neurobiology*, edited by S. B. Kater and C. Nicholson. Berlin: Springer-Verlag, 1973.
3. BARRETT, J. N., AND W. E. CRILL. Specific membrane properties of cat motoneurones. *J. Physiol. London* 239: 301, 1974.
4. BARRETT, J. N., AND W. E. CRILL. The influence of dendritic location and membrane properties on the effectiveness of synapses on cat motoneurones. *J. Physiol. London* 239: 325, 1974.
5. BLANKENSHIP, J. E. Action of tetrodotoxin on spinal motoneurons of the cat. *J. Neurophysiol.* 31: 186, 1968.
6. BLANKENSHIP, J. E., AND M. KUNO. Analysis of spontaneous subthreshold activity in spinal motoneurons of the cat. *J. Neurophysiol.* 31: 195, 1968.
7. BURKE, R. E., AND G. TEN BRUGGENCATE. Electrotonic characteristics of alpha motoneurones of varying size. *J. Physiol. London* 212: 1, 1971.
8. COLE, K. S. *Membranes, Ions and Impulses.* Berkeley: Univ. of California Press, 1968.
9. CONRADI, S. On motoneuron synaptology in adult cats. *Acta Physiol. Scand.* Suppl. 332, 1969.
10. COOMBS, J. S., J. C. ECCLES AND P. FATT. The electrical properties of the motoneurone membrane. *J. Physiol. London* 130: 291, 1955.
11. DEL CASTILLO, J., AND B. KATZ. The membrane change produced by the neuromuscular transmitter. *J. Physiol. London* 125: 546, 1954.
12. DIAMOND, J. The activation and distribution of GABA and L-glutamate receptors on goldfish Mauthner neurones: An analysis of dendritic remote inhibition. *J. Physiol. London* 194: 669, 1968.
13. ECCLES, J. C. *The Physiology of Synapses.* Berlin: Springer-Verlag, 1964.
14. GAGE, P. W., AND R. N. McBURNEY. Miniature end-plate currents and potentials generated by quanta of acetylcholine in glycerol-treated toad sartorius fibres. *J. Physiol. London* 226: 79, 1972.
15. GORMAN, A. L. F., AND M. MIROLLI. The geometrical factors determining the electrotonic properties of a molluscan neurone. *J. Physiol. London* 227: 35, 1972.
16. GRANIT, R. *Mechanisms regulating the discharge of motoneurons.* Springfield, Ill.: Thomas, 1972.
17. GRANIT, R., D. KERNELL AND Y. LAMARRE. Algebraical summation in synaptic activation of motoneurones firing within the 'primary range' to injected currents. *J. Physiol. London* 187: 379, 1966.
18. HODGKIN, A. L., AND W. A. H. RUSHTON. The electrical constants of a crustacean nerve fibre. *Proc. Royal Soc. London Ser. B* 133: 444, 1946.
19. HUBBARD, J. I. Mechanism of transmitter release. *Progr. Biophys. Mol. Biol.* 21: 33, 1970.
20. HUBBARD, J. I., R. LLINÁS AND D. M. J. QUASTEL. *Electrophysiological Analysis of Synaptic Transmission.* Baltimore: Williams & Wilkins, 1969.
21. IANSEK, R., AND S. J. REDMAN. The amplitude, time course and charge of unitary excitatory post-synaptic potentials evoked in spinal motoneurone dendrites. *J. Physiol. London* 234: 665, 1973.
22. JACK, J. J. B., S. MILLER, G. PORTER AND S. J. REDMAN. The time course of minimal excitatory post-synaptic potentials evoked in spinal motoneurones by Group Ia afferent fibres. *J. Physiol. London* 215: 353, 1971.
23. JACK, J. J. B., AND S. J. REDMAN. An electrical description of the motoneurone, and its application to the analysis of synaptic potentials. *J. Physiol. London* 215: 321, 1971.
24. KATZ, B. *The Release of Neural Transmitter Substances.* Liverpool: Liverpool Univ. Press, 1969.
25. KATZ, B., AND R. MILEDI. A study of spontaneous miniature potentials in spinal motoneurones. *J. Physiol. London* 168: 389, 1963.
26. KERNELL, D. High-frequency repetitive firing of cat lumbosacral motoneurons stimulated by long-lasting injected currents. *Acta Physiol. Scand.* 65: 74, 1965.
27. KERNELL, D. The limits of firing frequency in cat lumbosacral motoneurons possessing different time courses of afterhyperpolarization. *Acta Physiol. Scand.* 65: 87, 1965.

28. KERNELL, D. Input resistance, electrical excitability, and size of ventral horn cells in cat spinal cord. *Science* 152: 1637, 1966.

29. KRNJEVIĆ, K. Chemical nature of synaptic transmission in vertebrates. *Physiol. Rev.* 54: 418, 1974.

30. KUNO, M. Quantal components of excitatory synaptic potentials in spinal motoneurones. *J. Physiol. London* 175: 81, 1964.

31. KUNO, M. Quantum aspects of central and ganglionic synaptic transmission in vertebrates. *Physiol. Rev.* 51: 647, 1971.

32. KUNO, M., AND J. T. MIYAHARA. Nonlinear summation of unit synaptic potentials in spinal motoneurones of the cat. *J. Physiol. London* 201: 465, 1969.

33. KUNO, M., AND J. T. MIYAHARA. Analysis of synaptic efficacy in spinal motoneurones from 'quantum' aspects. *J. Physiol. London* 201: 479, 1969.

34. KUNO, M., AND J. N. WEAKLY. Quantal components of the inhibitory synaptic potential in spinal motoneurones of the cat. *J. Physiol. London* 224: 287, 1972.

35. LLINÁS, R., AND R. BAKER. A chloride-dependent inhibitory postsynaptic potential in cat trochlear motoneurons. *J. Neurophysiol.* 35: 484, 1972.

36. LLINÁS, R., R. BAKER AND W. PRECHT. Blockage of inhibition by ammonium acetate action on chloride pump in cat trochlear motoneurons. *J. Neurophysiol.* 37: 522, 1974.

37. LUX, H. D., C. LORACHER AND E. NEHER. The action of ammonium on postsynaptic inhibition of cat spinal motoneurons. *Exptl. Brain Res.* 11: 431, 1970.

38. LUX, H. D., P. SCHUBERT AND G. W. KREUTZBERG. Direct matching of morphological and electrophysiological data in cat spinal motoneurons. In: *Excitatory Synaptic Mechanisms.* Proc. of the Fifth International Meeting of Neurobiologists, edited by P. Anderson and J. K. S. Jansen. Oslo: Universitetsforlaget, 1970.

39. MAGLEBY, K. L., AND C. F. STEVENS. A quantitative description of end-plate currents. *J. Physiol. London* 223: 173, 1972.

40. PEACHEY, L. D. Transverse tubules in excitation-contraction coupling. *Federation Proc.* 24: 1124, 1965.

41. RALL, W. Branching dendritic trees and motoneuron membrane resistivity. *Exptl. Neurol.* 1: 491, 1959.

42. RALL, W. Distinguishing theoretical synaptic potentials computed for different soma-dendritic distributions of synaptic input. *J. Neurophysiol.* 30: 1138, 1967.

43. RALL, W. Time constants and electrotonic length of membrane cylinders and neurons. *Biophys. J.* 9: 1483, 1969.

44. RALL, W., AND J. RINZEL. Branch input resistance and steady attenuation for input to one branch of a dendritic neuron model. *Biophys. J.* 13: 648, 1973.

45. RINZEL, J. Voltage transients in neuronal dendritic trees. *Federation Proc.* 34: 1350, 1975.

46. WEAKLY, J. N. Effect of barbiturates on 'quantal' synaptic transmission in spinal motoneurones. *J. Physiol. London* 204: 63, 1969.

Index